ALL THUMBS

Guide to Telephones a
Answering Mac

Other All Thumbs Guides

Home Energy Savings
Home Plumbing
Home Wiring
Painting, Wallpapering, and Stenciling
Repairing Major Home Appliances
VCRs
Compact Disc Players
Home Computers
Fixing Furniture
Home Security
Car Care

ALL THUMBS Guide to Telephones and Answering Machines

Gene B. Williams

Illustrations by Patie Kay

TAB Books
Division of McGraw-Hill, Inc.
Blue Ridge Summit, PA 17294-0850

FIRST EDITION
FIRST PRINTING

Library of Congress Cataloging-in-Publication Data

Williams, Gene B.
Telephones and answering machines / by Gene B. Williams.
p. cm.
Includes index.
ISBN 0-8306-4435-0 (paper)
1. Telephone—Equipment and supplies—Amateurs' manuals.
2. Telephone answering and recording apparatus—Amateurs' manuals.
I. Title.
TK9951.W55 1993
621.386—dc20
93-12882
CIP

Acquisitions editor: Roland Phelps
Editorial team: Melanie D. Brewer, Editor
Robert Ostrander, Executive Editor
Stacey R. Spurlock, Indexer
Production team: Katherine G. Brown, Director
Brenda S. Wilhide , Layout
Kelly S. Christman, Proofreading
Joyce Bellela, Computer Graphic Artist
Design team: Jaclyn J. Boone, Designer
Brian Allison, Associate Designer
Cover design: Lori E. Schlosser
Cover illustration: Denny Bond, East Petersburg, Pa.
Cartoon caricature: Michael Malle, Pittsburgh, Pa.
ATS
4416

The All Thumbs Guarantee

TAB Books/McGraw-Hill guarantees that you will be able to follow every step of each project in this book, from beginning to end, or you will receive your money back. If you are unable to follow the All Thumbs steps, return this book, your store receipt, and a brief explanation to:

All Thumbs
P.O. Box 581
Blue Ridge Summit, PA 17214-9998

About the Binding

This and every All Thumbs book has a special lay-flat binding. To take full advantage of this binding, open the book to any page and run your finger along the spine, pressing down as you do so; the book will stay open at the page you've selected.

The lay-flat binding is designed to withstand constant use. Unlike regular book bindings, the spine will not weaken or crack when you press down on the spine to keep the book open.

Dedication

To all the graduates of Central High School in Minneapolis. The school is no more (although I have one of its bricks), but it won't be forgotten.

Acknowledgments

My thanks to everyone who made this book possible. Melanie Brewer and Roland Phelps handled much of the drudgery. My artist, Patie Kay, spent countless hours making sure that the illustrations were helpful and also that the book, in general, made sense. My young son, Danny, was of more help than he knew. He would pop in at just the right moment, look at one of the drawings of a ripped-apart telephone and ask "What's that?" Sometimes in explaining it to him, we came up with a better explanation for you.

The most important people in this project are you, the readers. More than anyone else, you have helped to make this book series a success.

Thanks to all of you!

Contents

Preface

A collection of books about do-it-yourself maintenance and repairs, the All Thumbs series was created not for the skilled jack-of-all-trades, but for the average person whose budget isn't keeping pace with today's rising costs. If your familiarity with the various systems in the home is minimal, this series is tailor-made for you.

Because service calls are time-consuming, often requiring more travel time than actual labor, they can be expensive. The All Thumbs series saves you time and money by showing you how to make most common repairs yourself. Just as important, these handy guides walk you through the cleaning procedures that will keep your telephone and answering machine, as well as other electronic systems in your home, up and running for years.

The guides cover topics including home computers, VCRs, CD players, as well as car care; home wiring; plumbing; painting, stenciling, and wallpapering; and home energy savings to name a few. Copiously illustrated, each book details the procedures in an easy-to-follow, step-by-step format, making many repairs and improvements well within the ability of the average reader.

Introduction

There was a time when getting a message from one place to another meant writing it down and sending it with someone on foot or horseback. Eventually came the telegraph, which could send coded messages across wires. It was a fairly short step from there to sending voices the same way.

From here it got complicated. Most of this end is never noticed by the end user until something goes wrong. In essence, the telephone system has your telephone connected to every other telephone. At least in America (and other places), all this happens with remarkable reliability.

Even so, things can happen. It may be with the lines coming to your home, with the lines inside the walls or with the telephones and other devices themselves. With changes in the various telephone companies, you are usually expected to take care of at least the last part yourself. In fact, if it turns out that the whole problem is with one of your phones, you can be charged a hefty fee for the service call.

It has become important, and prudent, to learn at least some basics about your home telephone system. *That* is what this book is all about.

Chapter 1 explains the importance of safety. The tools you'll need and how to use them are described in chapter 2. Chapter 3 explains the basics of how it all works, which is important if you are to do any kind of troubleshooting. Troubleshooting tips are given in chapter 4.

A number of things can happen to the wiring and jacks. In chapter 5 you'll learn about the wiring, and chapter 6 tells you about how this wiring is connected. Between the two you'll discover how to check this very important part of the telephone system, how to make repairs to it, and even how to install a new jack.

It's not unusual for a telephone to last for years or even decades. Even so, things can go wrong. Chapter 7 shows you the different kinds of telephones and how you can quickly test, and often repair, them. Chapter 8 covers a special kind of phone—the cordless. Chapter 9 completes the "meat" of the book with information on answering machines.

The most complicated tool called for in this book is the volt-ohmmeter (VOM). At the end of the book is an appendix on how to buy and use this tool. (It's easy). A second appendix is provided to walk you through two ways for making a simple but functional signal generator so that you can test the wires inside the walls of your home.

The most important thing you'll learn is that you have fewer limitations than you might think. More often than not, you can diagnose at least the cause of the problem in a matter of minutes. With the information in this book you'll be able to take care of nearly every problem.

CHAPTER ONE

Safety

Safety is essential for you to know and practice. It is important for everyone, especially for beginners who might not know all of the potential dangers. Fortunately there are very few hazards in and around telephones.

A telephone that is idle presents no danger to you. Low voltage might be present on the wires, but it is not at a dangerous level. However, if someone calls in while you are touching the wires, you can get a nasty shock. The amperage is low, but the voltage is high. You're not likely to be permanently injured, but it will hurt!

An even greater danger is present during a storm that has lightning. In theory, the lines are protected. Don't trust your life to this. If there is a storm, you shouldn't even be talking on the phone let alone working on the wiring.

Some telephone-related equipment (answering machines, cordless phones, etc.) are plugged into a standard wall outlet. In all cases the power coming out of the wall outlet and going into the power supply of a device can measure up to 120 lethal volts.

Getting a jolt from the incoming 120 volts ac (120 Vac) is more than just unpleasant; it can be fatal. Studies have shown that it takes very little current to kill. Even a small amount of current can paralyze your muscles and you won't be able to let go. Just a fraction more and your heart muscle can become paralyzed.

This equipment is designed specifically to be safe. You have to take some purposeful steps to get anywhere near spots that might be dangerous. The exception is when a power cord becomes frayed.

Now that you are sufficiently frightened, relax. As long as you are careful, you take your time, and you follow rules of safety and common sense, you won't get into trouble. Learn where the dangerous spots are, stay away from them, and you'll have no problems. You're more likely to cause damage to the phone or related equipment than to yourself.

Step 1-1. Finding the holding screws and clips.
It's not always obvious how a telephone or other device is held together. The design usually calls for cosmetic attractiveness. Holding screws tend to be hidden under the unit rather than in a visible place. In the illustration, the screw that holds the handset together is "invisible" beneath the paper and plastic cap that is used to display the phone number.

If you need to get inside and the casing doesn't come apart easily, do not force it open. This rule is true regardless of the device.

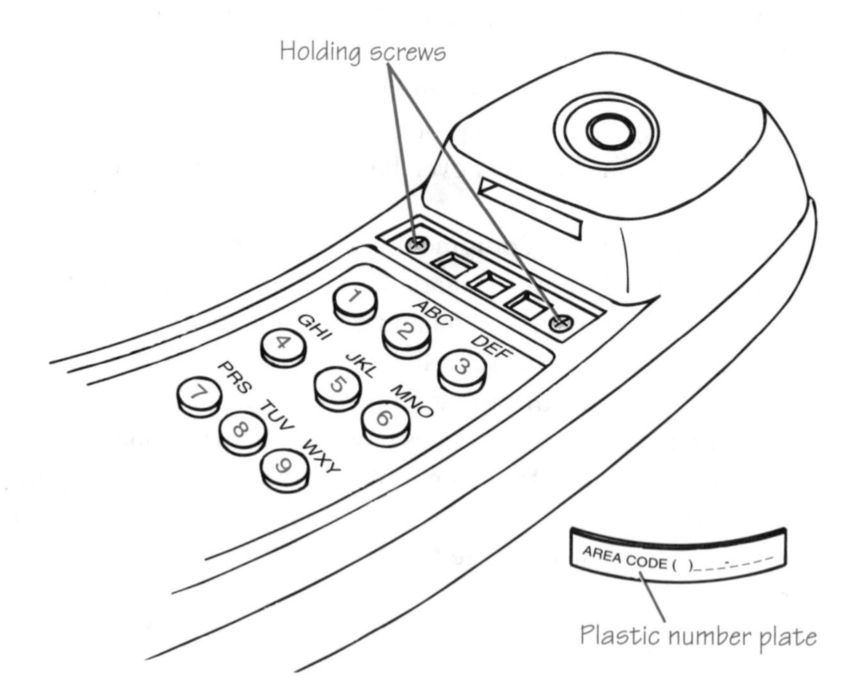

Step 1-2. Locating prying slots.
Sometimes a device, particularly a handset, will be held together by a holding screw and also by catches. In almost every such case, there will also be prying slots. Keep in mind that the plastic is fairly soft and can be easily damaged. Proceed cautiously.

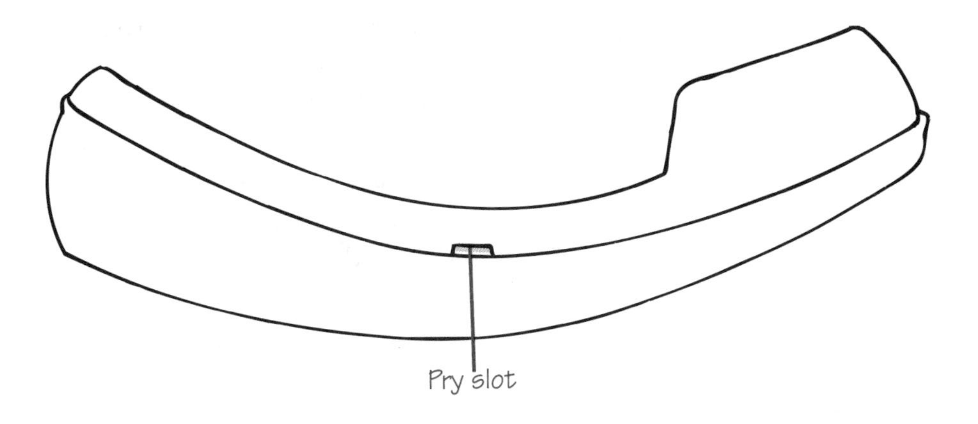

Step 1-3: Disconnecting the incoming lines.
If you intend to work on the wires, it's often a good idea to disconnect the home from the incoming lines at the access box. This protects you, and the test equipment, from the high voltage surge present if someone calls in. It's critical that you note the color coding on the access box lugs.

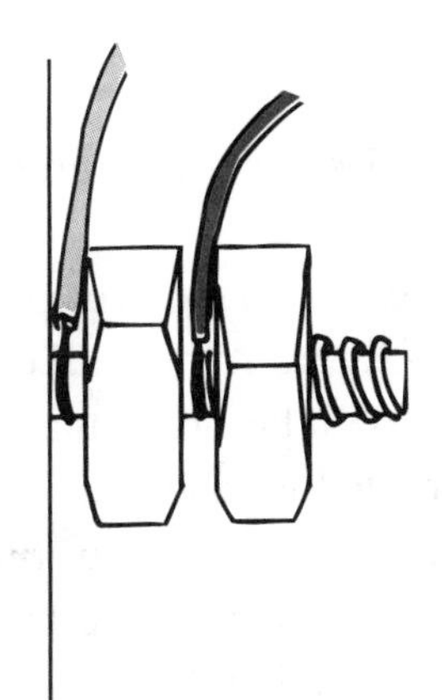

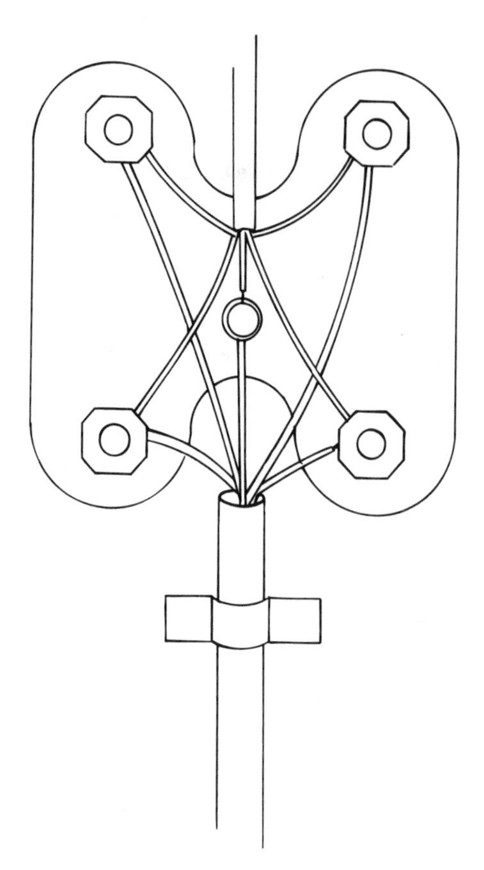

If these lugs are not already color coded (red, black, yellow, green), take a moment to put some kind of marking on them yourself before disconnecting the wires. Remember that there could be dangerous voltage present while you make the disconnect. Insulate yourself from the surroundings.

Step 1-4. Unplugging the device.
If the device receives power directly from the wall outlet, unplug it before you attempt to remove the case and get inside. The correct way to unplug any appliance is to grasp the plug. Never pull on the cord, which can cause it to pull loose from the plug.

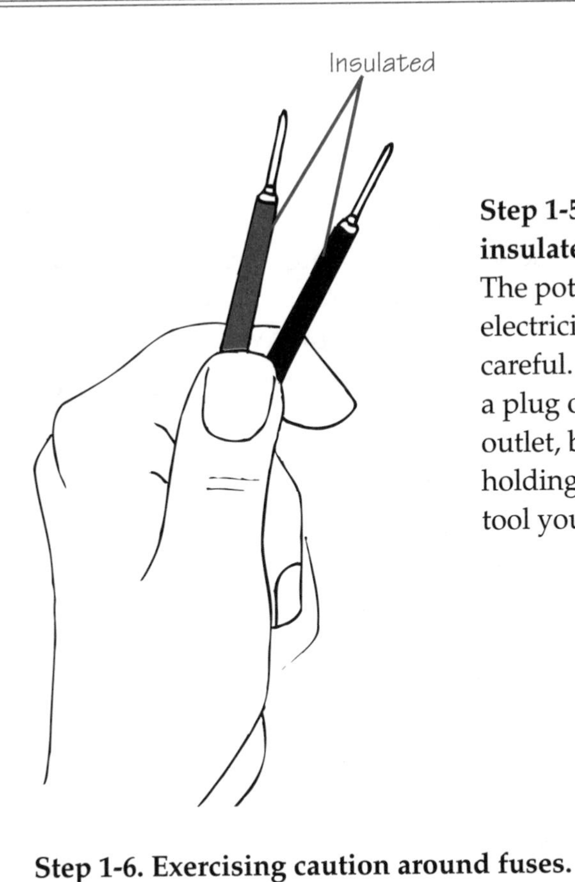

Step 1-5. Touching only what is insulated.
The potential danger from electricity begins at the outlet. Be careful. Any time you are inserting a plug or a VOM probe into the outlet, be sure that your fingers are holding the insulated part of the tool you are using.

Step 1-6. Exercising caution around fuses.
Devices powered directly by connection to a wall outlet will have an internal power supply. Most often these devices will have a fuse or fuses. Be careful around fuses. Almost all fuses are used to protect the computer from 120 volts ac. These fuses are located near the power supply. As long as the unit is plugged in, there is danger present.

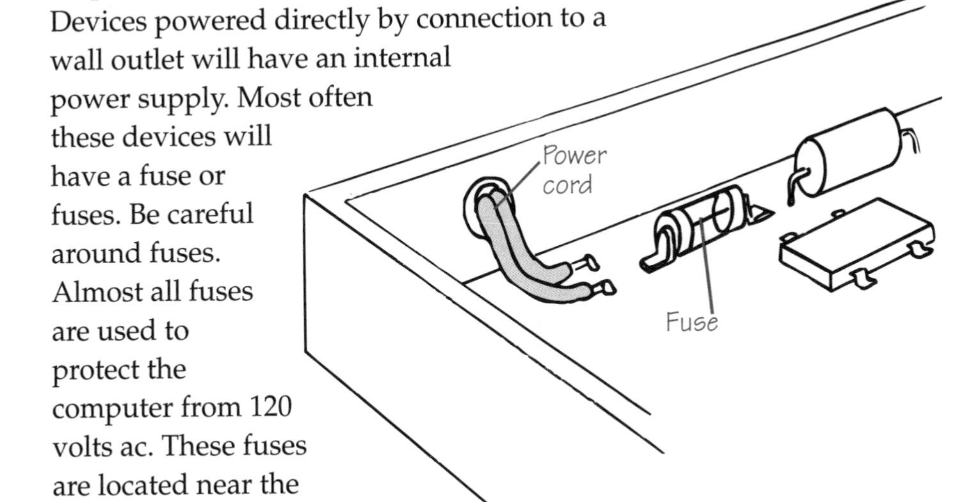

Step 1-7. Moving slowly and carefully.
Metal probes and tools can cause short circuits and other damage. There is no danger if you've unplugged the device you are working on, but there are times when power will have to be present for you to proceed. At such times, paying attention to what you are doing becomes more important than ever. When working inside an electrical component, move slowly and carefully. It's easy to cause damage to the delicate parts. The more electronic the phone or device is, the more likely it is to cause damage.

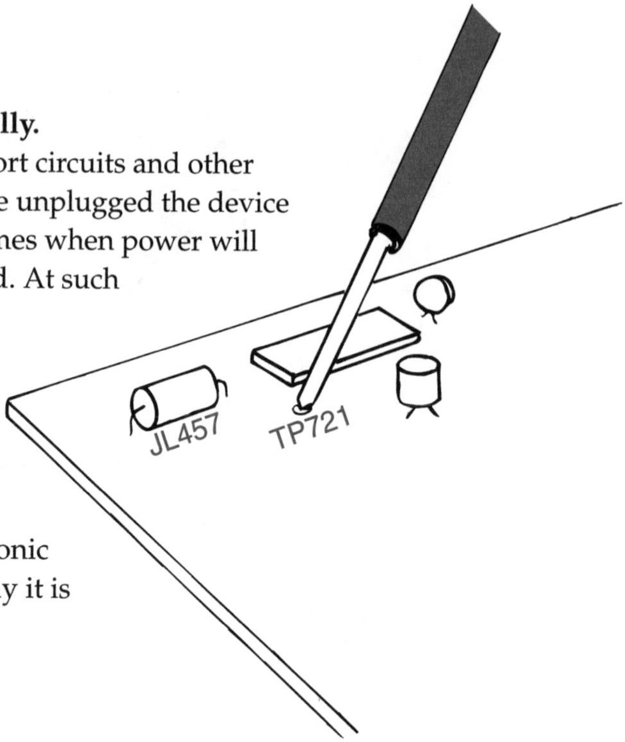

CHAPTER TWO

Tools

You probably have most of the tools you'll need around the house. None of the tools and materials needed in this guide are costly. The most expensive is the optional volt-ohmmeter (VOM). For the purposes of this book, a VOM unit costing about $10 to $20 is sufficient.

Avoid buying poor-quality tools. A cheap screwdriver might do the job just as well as an expensive one; however, it also can cause you considerable trouble. There are reasons why a cheap tool is cheap. The materials used are of lower standards. The metal can bend and break, or the coating can flake off and fall inside the machine where it can do considerable damage.

Some of the tools listed in this chapter are optional and don't need to be purchased until and unless you need them. For example, there is no need to get a set a nut drivers if nothing you own has hex-head screws or bolts. You may or may not need any of the cleaning materials. In short, you'll be surprised at what you can do with nothing more than a standard screwdriver.

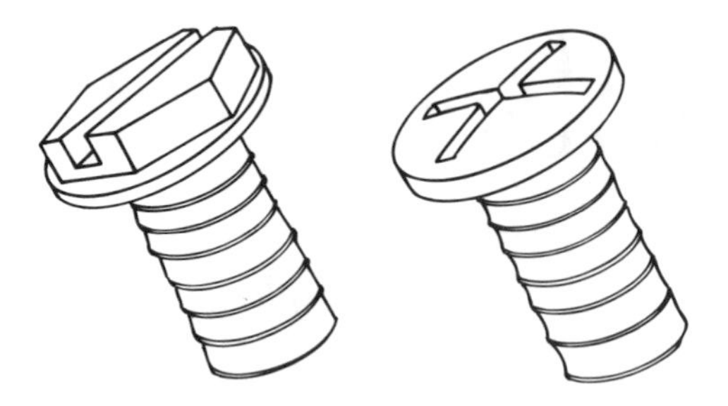

Screw types

Telephones and related devices are usually held together with screws. These either have a slot or an "x." The slotted type calls for a blade-type screwdriver; the other type requires a Phillips-type screwdriver.

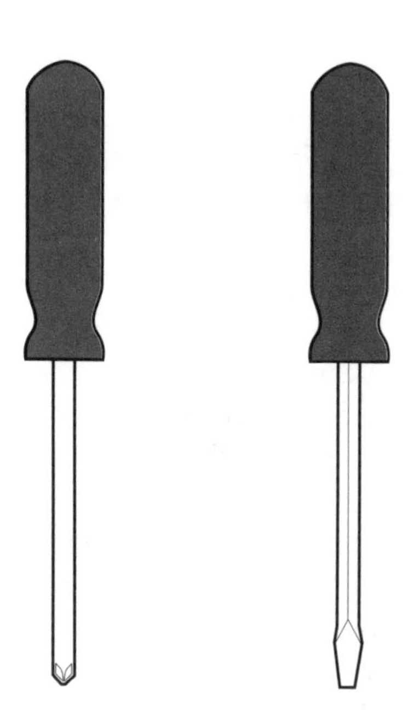

Screwdrivers

In both cases, the most versatile screwdrivers are those with a medium-sized head (about 1/4 inch) and those with a small-sized head (about 1/8 inch). The shaft length isn't important, but most people find a length of 4 to 6 inches the most convenient and versatile.

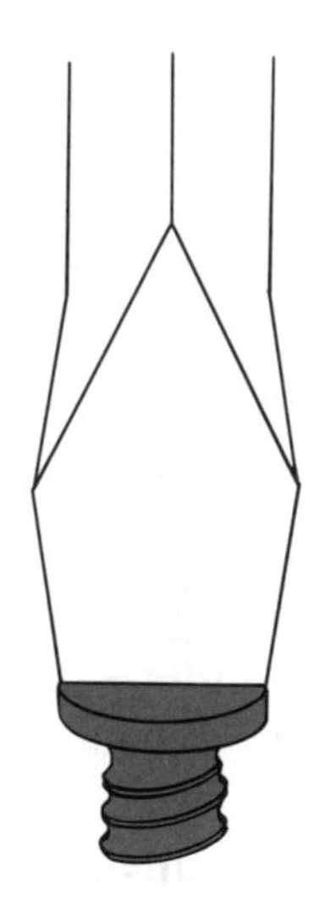

Fit the blade to the screw

It's important that the screwdriver fits the screw, both for the length of the slot and for the width. The same is true of the Phillips screw and screwdriver. The head should fit the slot.

Nutdrivers

Some equipment, and some access boxes, are held by bolts or screws with hex heads. A box or adjustable wrench *can* be used, but the nutdriver makes the job much easier. This is like a screwdriver handle with a socket attached to the end. In any case, the fit is critical. Nutdrivers usually come in complete sets. English units (inches) should be sufficient.

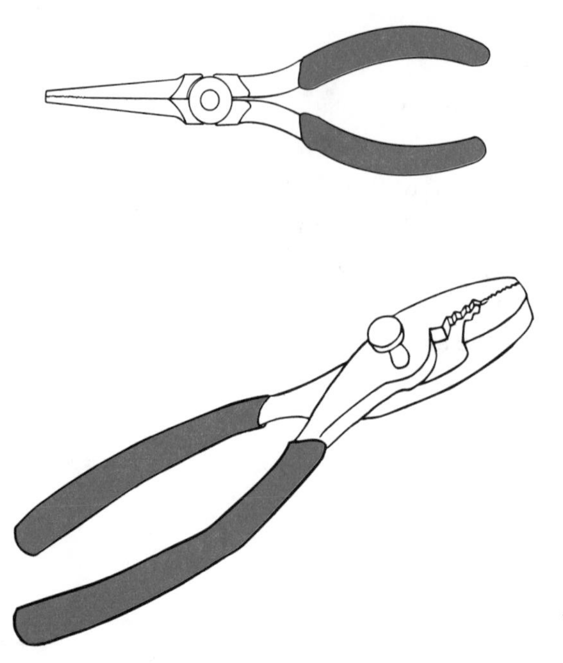

Needle-nose and standard pliers

Needle-nose pliers are good for light gripping, parts retrieval, and various other tasks. A standard pair of pliers also might come in handy. In both cases, be sure that the handles are insulated.

Isopropyl alcohol for cleaning

Especially if you intend to do any maintenance on an answering machine, you will need a fluid for cleaning certain parts. Isopropyl alcohol is the standard cleaning fluid. It's inexpensive and relatively safe to use. Be sure to get technical grade isopropyl alcohol with a purity of at least 95 percent. ***Caution!*** Do not use alcohol on any rubber parts (the rollers inside an answer machine, for example).

Special cleaning fluids

Special cleaning fluids are more expensive than alcohol; however, they are often better cleaners. Get a cleaner that leaves no residue and is safe for rubber parts. Usually the label on the container will tell you if the product is a nonresidue solvent.

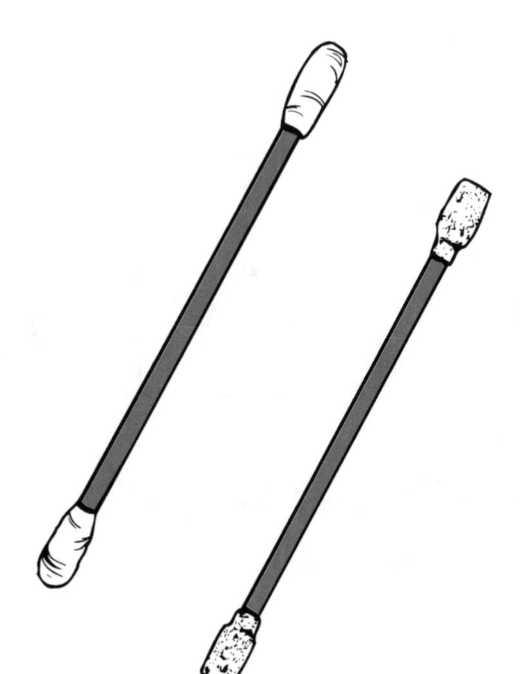

Cleaning swabs

The swabs used must be of the appropriate material. Those with threads, like cotton swabs, are tolerable for general inside cleaning. More expensive but better are the foam swabs found in most electronics supply stores.

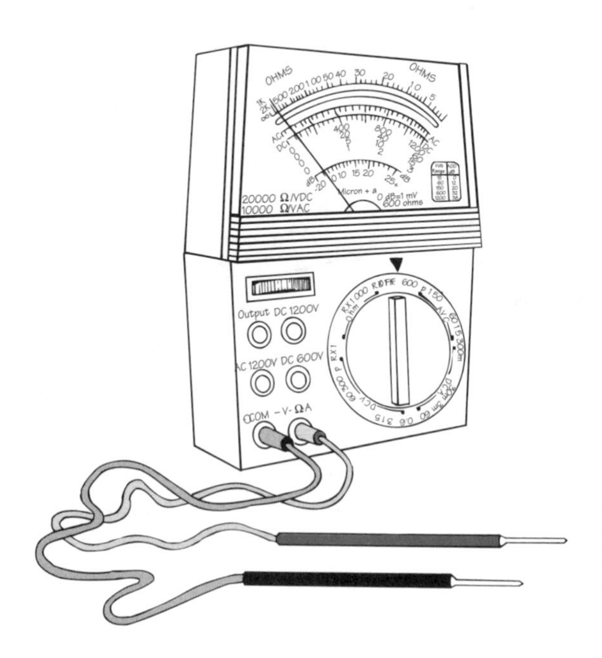

VOM

A VOM (volt-ohmmeter, also called a multimeter) is used to measure voltage and resistance. It is an invaluable tool to have around the house because you can make power tests, continuity tests, and other tests. Most of the tests you'll be making will not require a high degree of accuracy. (See appendix A.)

Signal generator

The VOM is great for testing for voltage and continuity when the two test points are close. For testing one of the wires in the wall you will need a way to send a signal through the wire. Appendix B shows you an easy way to make and use your own *signal generator*. Don't let the name intimidate you. A signal generator is anything that can be used to provide a predictable signal. It can be something as simple as a battery.

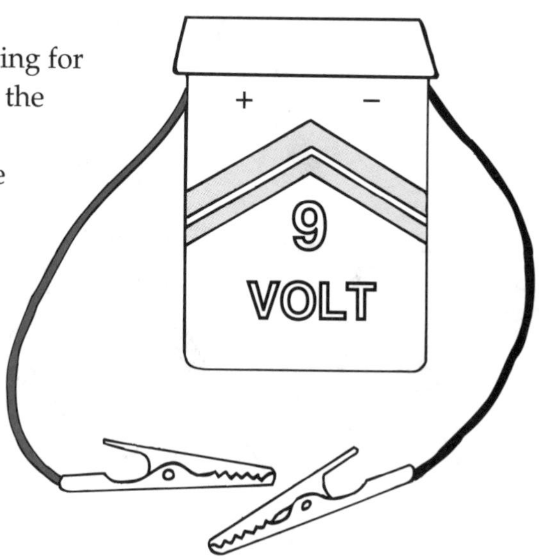

A telephone

One of the most important pieces of test equipment is a phone you know to be in good working condition. This will tell you instantly if the problem is being caused by a telephone or by the wiring.

Caulking

Most caulking is available in a squeezable tube. Instructions for use are on the tube. The two most common uses are to seal holes and to hide wires, such as those routed beneath the home's eave.

Stapler

To hold the wires in place, you might need to use a staple gun. This tool must be large and strong enough to insert the staples into the wood. An office-type stapler won't work.

The staples are usually of the chisel-point type. Use great care to not cut, crimp, or damage the cable.

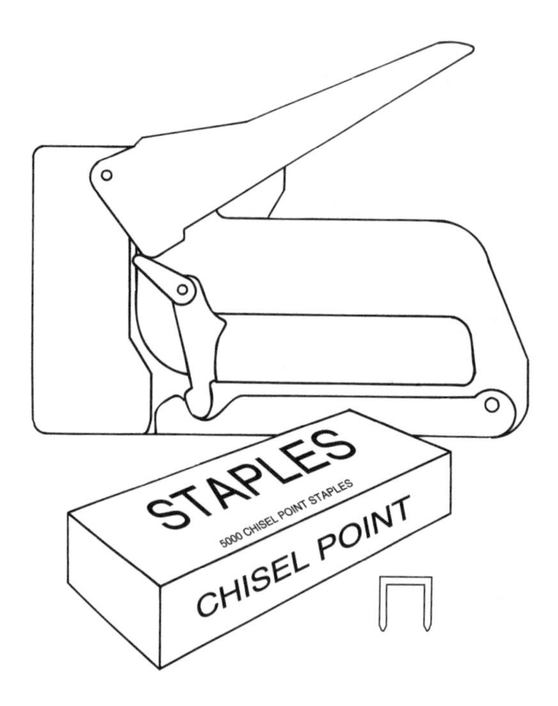

Clamps

Sometimes it is more convenient and practical to use a clamp to hold the telephone cable. Some nail into place. Others use a screw.

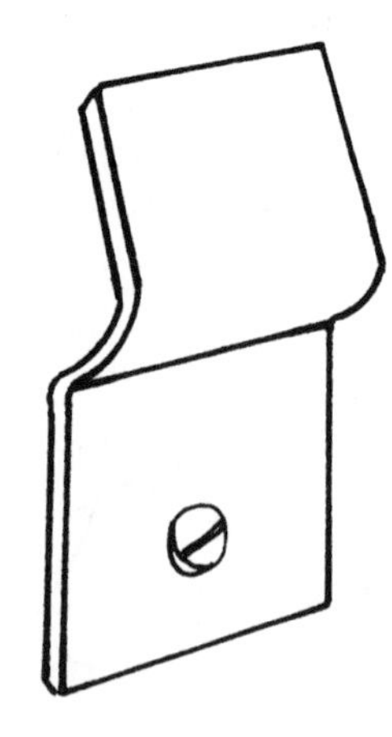

Drill

An electric drill can be used to bore holes through walls or beams so that the telephone wires can be inserted. In most cases, a standard 1/4-inch drill can be used. For certain jobs, however, a special drill bit will be needed (for example, you would use a masonry bit to drill through brick or concrete, or a bit with a long shaft to drill through a thick wall).

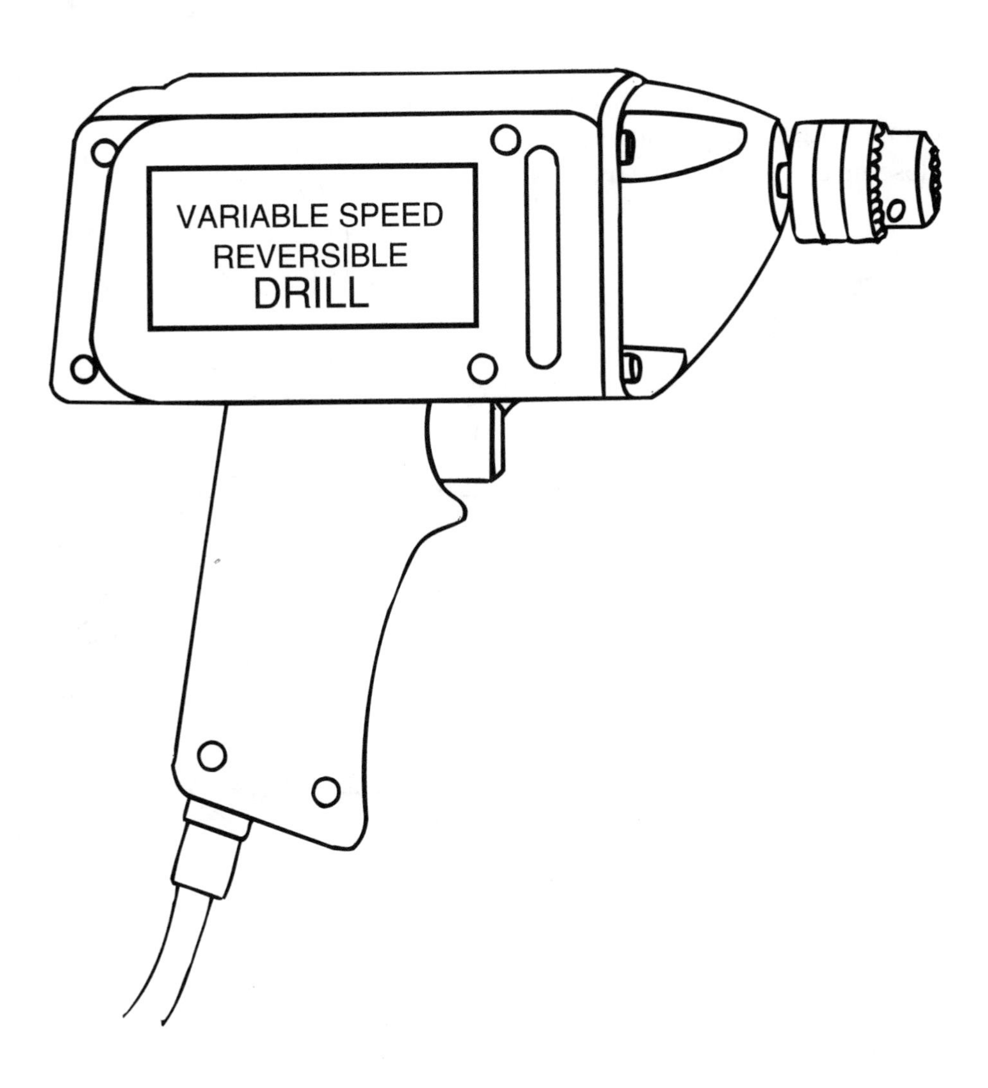

Miscellaneous household items

Many of the miscellaneous things you'll need can be found around the house. A muffin tin or paper cups can be used to hold removed screws or other parts. A flashlight can help you see into dark places inside the component. A magnifying glass is good for close-up examination of small parts.

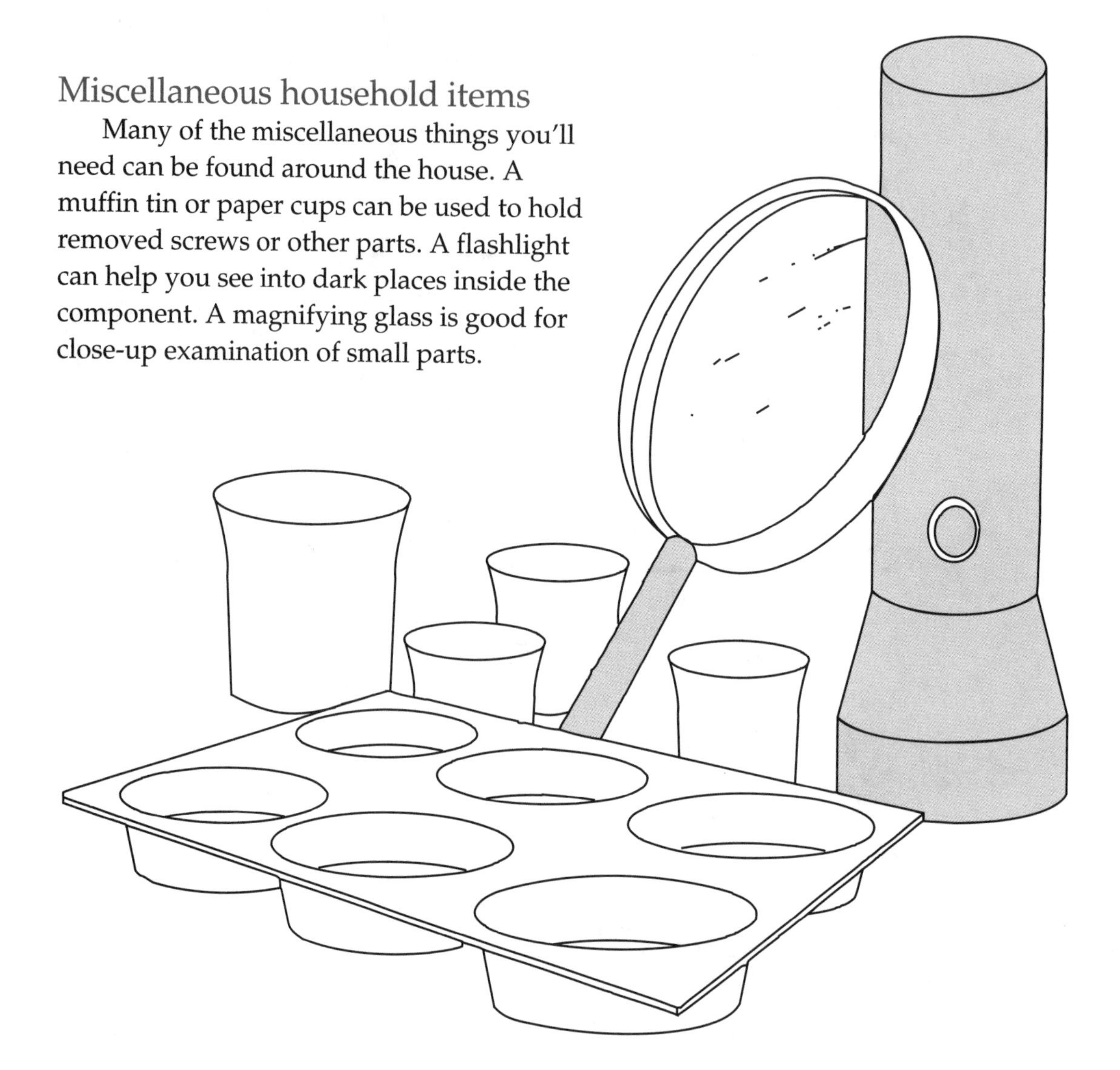

CHAPTER THREE

How Your Telephone System Works

Imagine trying to fix a car without having any idea about what makes it work. You generally don't need this understanding if you are merely operating the device or machine, but the knowledge is essential when it comes to troubleshooting and repairing a device.

Your goal isn't to learn how to design and engineer a telephone system or any part of it. All you need to understand are the basics of how it operates. With the restructuring of phone companies across the country and elsewhere, knowing these basics can mean the difference between making a call or paying a large fee just to find out that the battery in your cordless phone is causing the annoying static you were complaining about.

Your greatest enemy is "high-tech intimidation." Being cautious and recognizing your honest limitations is wise. Being intimidated to the point that you feel totally incompetent isn't necessary.

The overall phone system

The phone company's central (switching) office contains the computerized automatic switching. How the call is routed is determined by the numbers this computer receives. A main cable, or trunk, from the central office comes to a pedestal. All the phones in the local area are connected to the trunk line here.

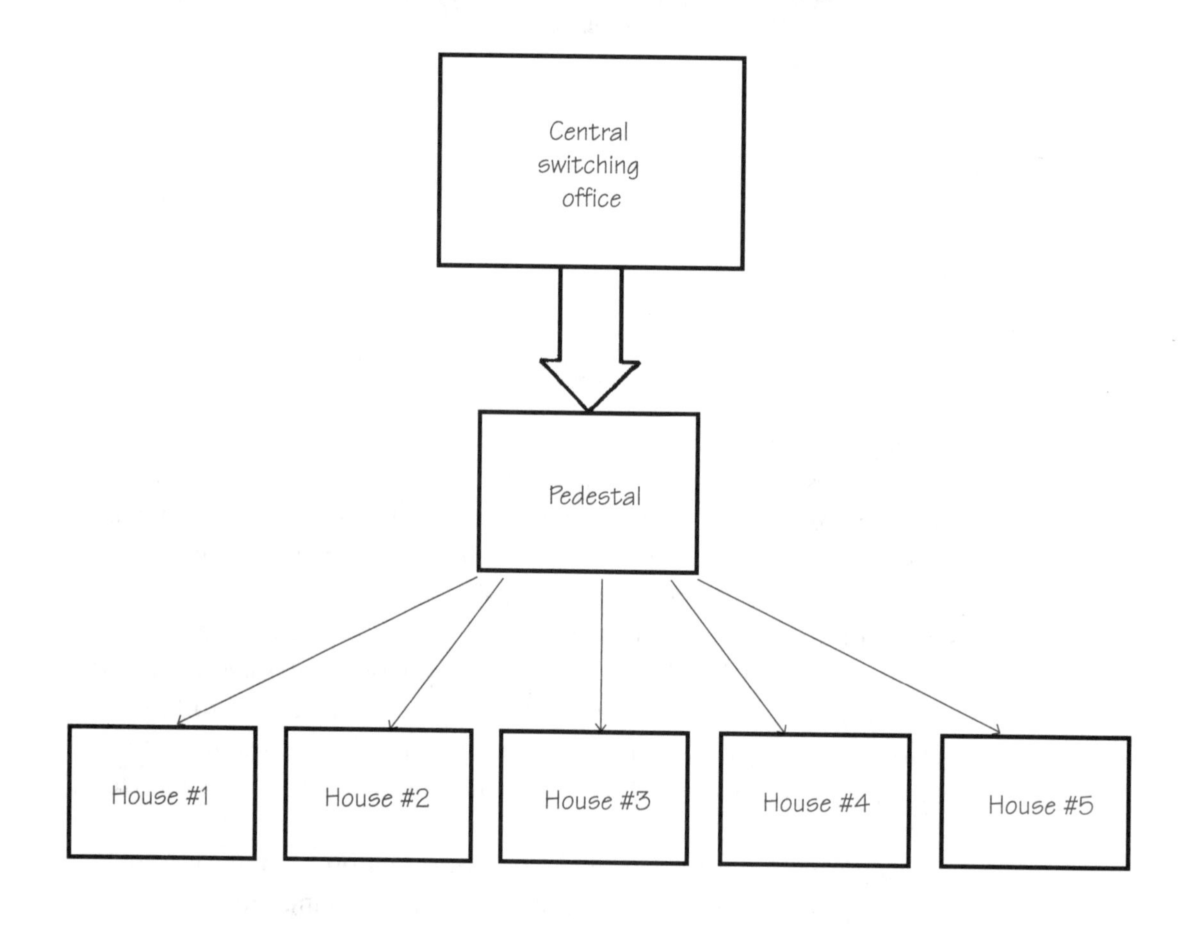

The wiring in the house

Most often, the cable brought to the house will have two pairs of wires inside. These wires will be red, black, yellow, and green. The cable is attached to the access (service) box. In many cases these days, this is where the phones company's responsibility ends. (More detail on home wiring is given in chapters 5 and 6.)

From the access box, the wires run through the house, often inside the walls, to each phone outlet. This can be done in series to save wire and work. The disadvantage of this method of wiring is that if any jack goes bad, the entire system beyond it will seem to be bad.

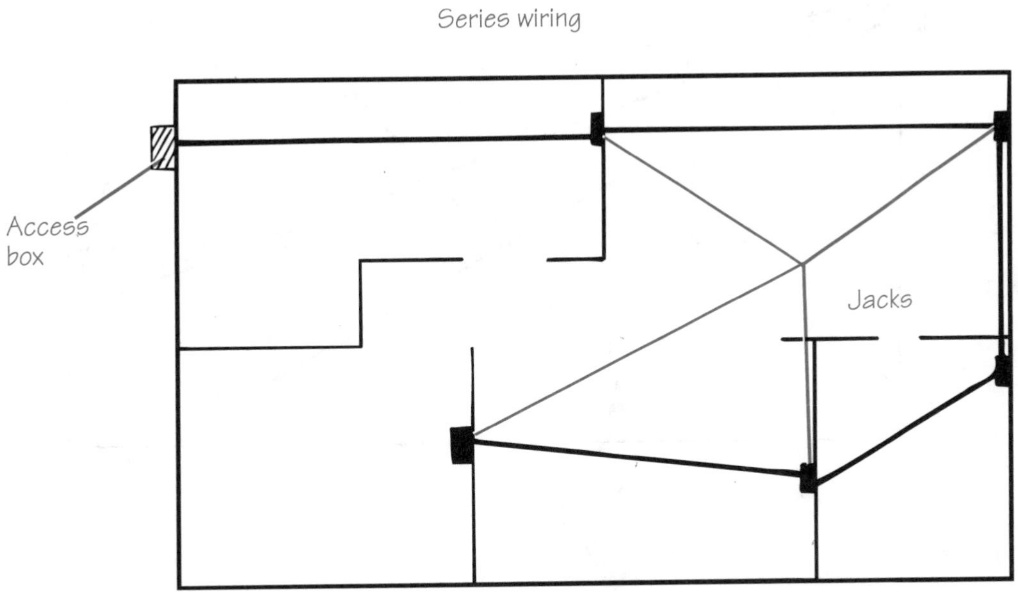

Series wiring

Access box

Jacks

Generally preferred is parallel or "home run" wiring. With this method, each jack has its own cable, with all of the wire running back to a junction box at the access.

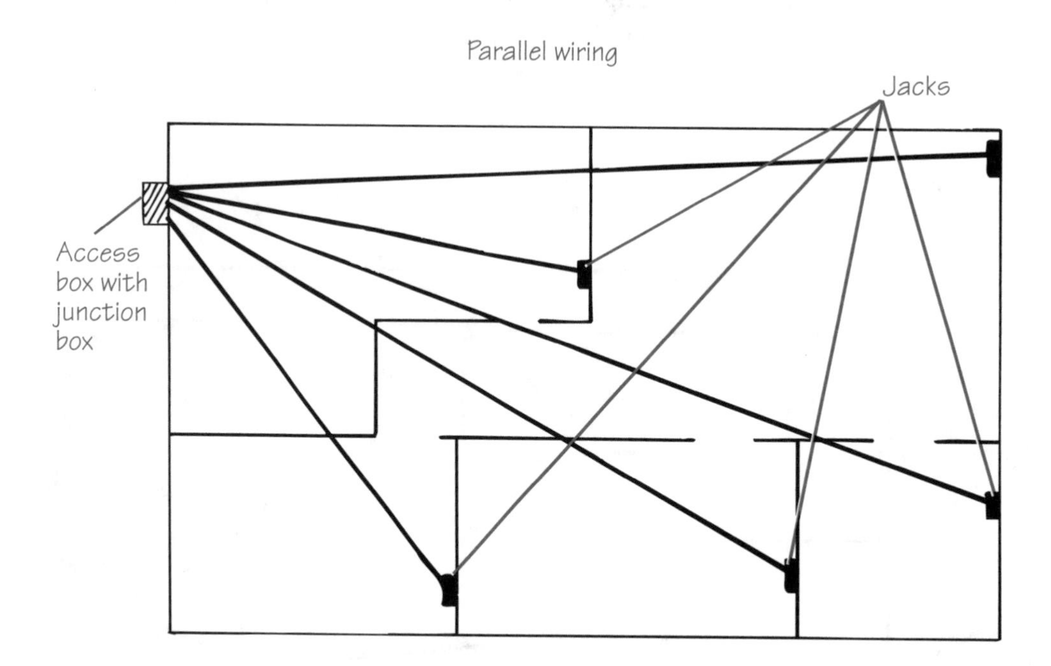

The wire pairs

A telephone needs only two wires to operate. This is "the pair" and they are usually color coded. The most popular combinations are red-green and yellow-black. This means that most often the telephone will be connected to the red and green wires. If these are already in use, or have gone bad, the yellow and black wires will be used. Your own wiring might be different, which is fine. The only important factor is that the same pair be used the same way throughout the system.

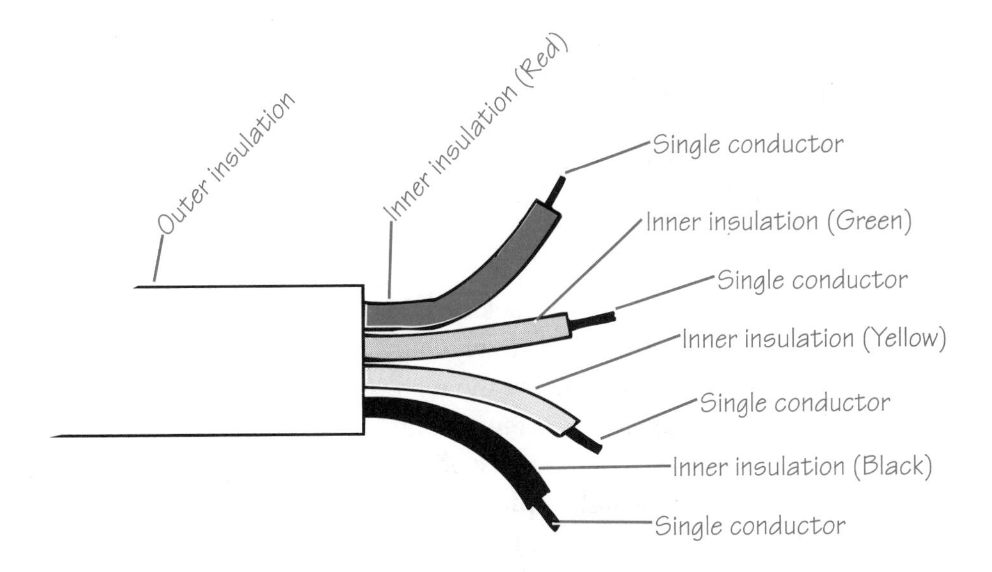

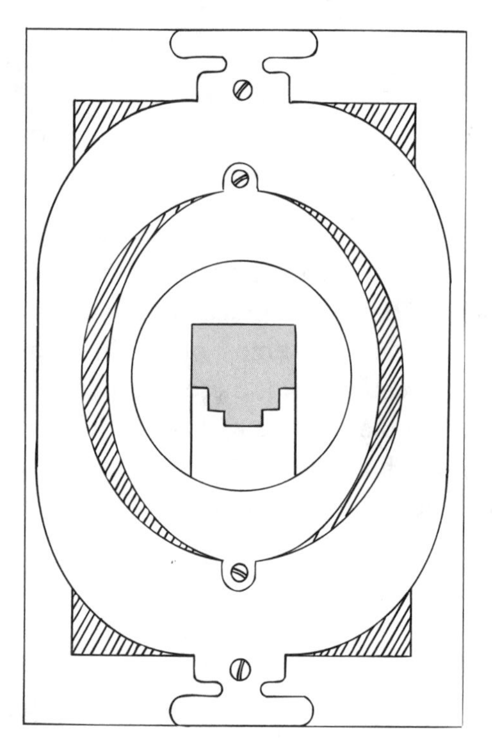

The jack

Jacks and connectors are covered in detail in chapter 6. There are several kinds of jacks, but all of them are similar in that there are screws inside where the incoming wires connect.

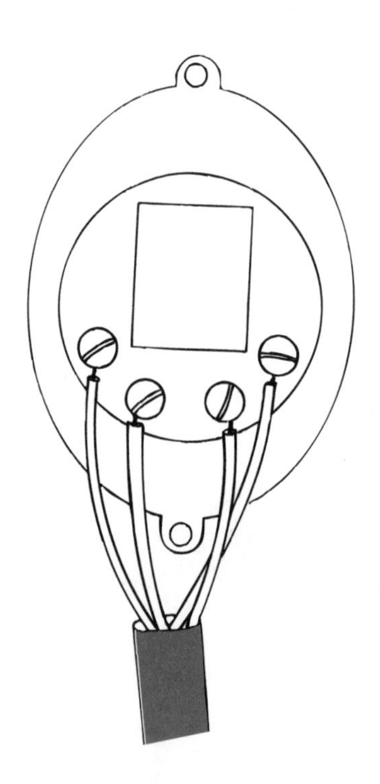

The standard wall jack consists of a plastic or metal box, which is attached to a stud inside the wall. A back or base plate is screwed onto this box. The wires usually pass through a hole in this plate, although they may connect in the back. The wires are connected to the jack in the plate. The installation is completed with a cover plate.

The microphone

As you speak into the phone, air vibrations caused by your voice cause a diaphragm in the mouthpiece to move. This generates electricity that modulates (changes) with your voice.

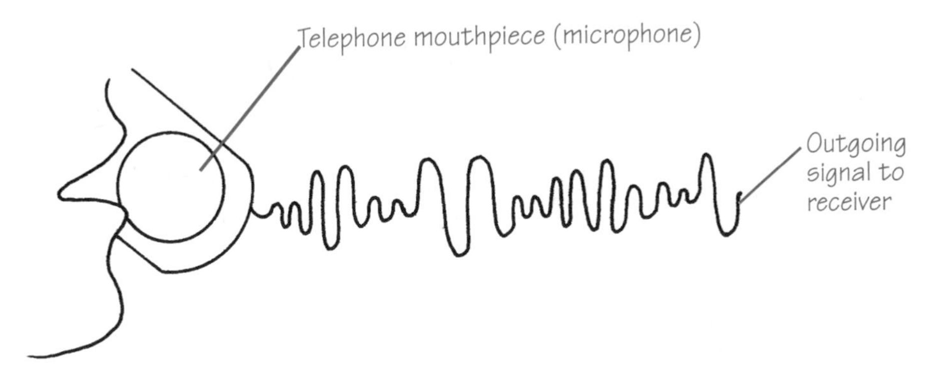

The earphone

In the earphone the opposite happens. The incoming voice is carried as a modulated electrical current, with the frequency being "in step" with the speaker's voice. This causes a diaphragm to be moved, which vibrates the air and makes the sound you hear.

Incoming signal

Telephone receiver

Pulse dialing

Some phones, particularly older models, have a round dial. As you dial a telephone number, electrical pulses are sent. The farther the dial is rotated, the more pulses there are. The telephone company's equipment "counts" these pulses.

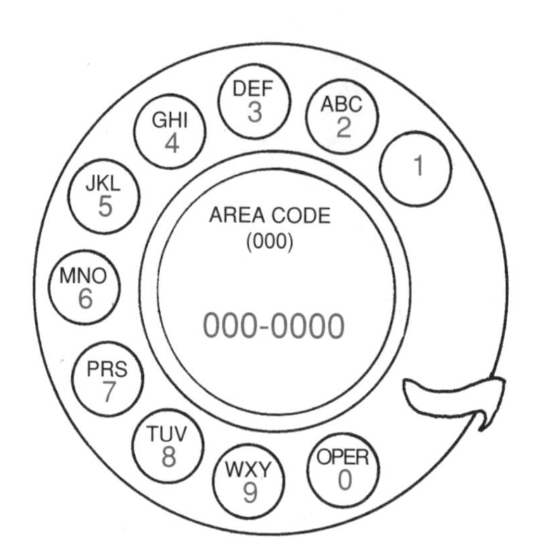

Tone dialing

More and more phones use "touch tone" dialing now. A signal (tone) generator in the touch pad generates different tone combinations for each key. All the keys on a particular vertical will have the same tone, as will all the keys on each horizontal line. When a particular key is pressed, both tones are generated. It's the combination of tones that is unique to each key. This is sometimes referred to as DTMF, for *d*ual *t*one *m*odulating *f*requencies.

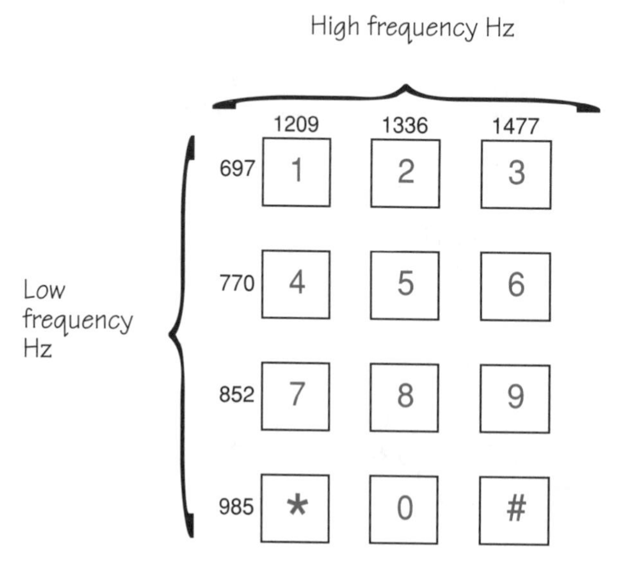

CHAPTER FOUR

Basic Troubleshooting

Troubleshooting is a step-by-step process of elimination. Once you've eliminated all the places where the problem is not, you can find out where it is. Some examples of common problems are given in this chapter. At times the procedures for solving common problems require tests on wires. *Be careful!* (Refer to chapter 1.) There is voltage present and this voltage can be dangerous if someone calls in while you are touching the wires. If you don't feel competent to do something that might be dangerous to you, don't do it! Call a professional technician.

The process of troubleshooting always begins with the simple and progresses to the more complex. Start by looking for the obvious. Where you begin is determined by the general symptoms your system is generating. For example, if one phone is completely dead, but other phones in the home are working fine, the problem will not involve the lines outside the home. Even if your phone company offers a service policy on in-home wiring, you could still be charged for the service visit if the fault is the dead telephone.

Ask yourself some questions.

Has it ever worked? You might be trying a new telephone, or might be trying to get it to do something new (such as using the memory on an autodial phone). It's possible that you're doing something wrong, or that you're trying to do something that can't be done.

Next, does it work at all? If the phone is receiving calls fine but won't dial out, the wiring in the home is probably okay.

Although this chapter is given in a step-by-step format, the actual beginning point is determined by what is happening, and what is not happening. Wherever you begin, proceed from one step to the next.

This chapter covers the system as a whole. For specific information, turn to the appropriate chapter. Chapter 7 covers the telephone itself, chapter 8 deals with cordless phones, and chapter 9 tells you more about troubleshooting answering machines.

Step 4-1. Making preliminary checks.

Troubleshooting always begins by checking the easy and the obvious. If nothing at all is happening, look to see if the unit is plugged in, both into the wall jack and into the phone.

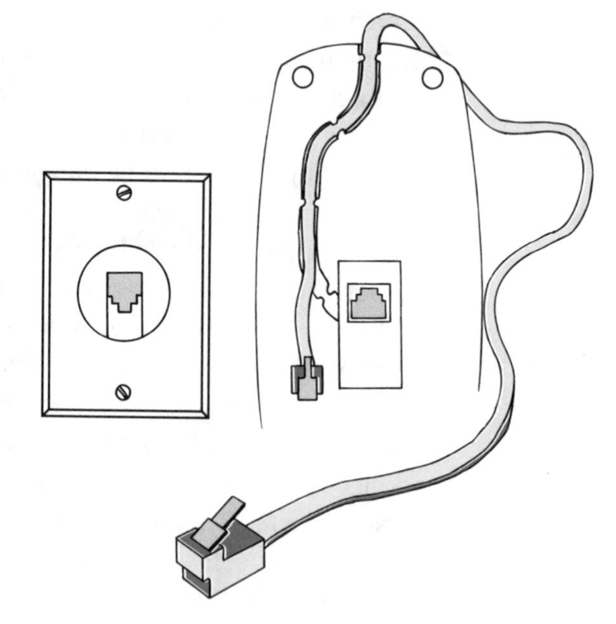

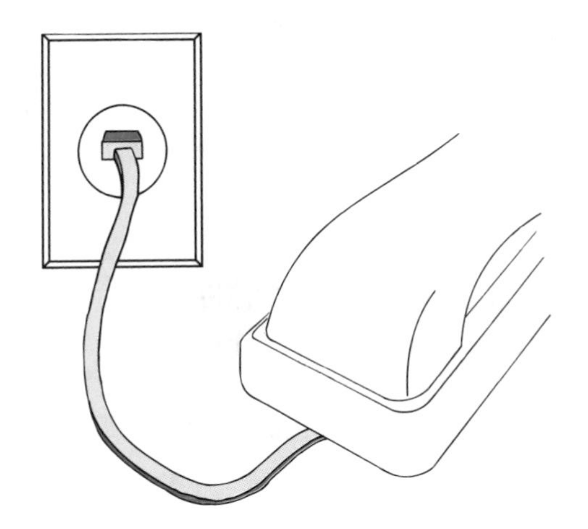

Plug another phone into the suspected outlet. It's possible, but unlikely, that both phones will begin to malfunction at the same time. (To be sure, you can take one of the phones to the home of a friend or neighbor and plug it in. If it works, you know that the problem is not in the phone itself.)

The testing can continue by bringing both phones to another outlet, preferably one closer to the access box. If you want, you can test every outlet in the house this way. If more than one outlet is dead there are only three possibilities. One possibility is that something is wrong with the line coming to the house, which means that the phone company must be called. Another possibility is that your wall jacks have been wired in series (see chapter 5) and the damaged wire is closer to the access box. The third possibility is extremely unlikely. It would mean that a parallel ("home run") wiring is being used but every cable in the house has been simultaneously damaged.

It's possible that one phone or device is shorted and is "bringing down" the rest of the system. Go through the home and unplug every phone and device but a phone you know is good. If the system functions now, one of the devices is bad. You then need to connect them again, one at a time, until the system fails.

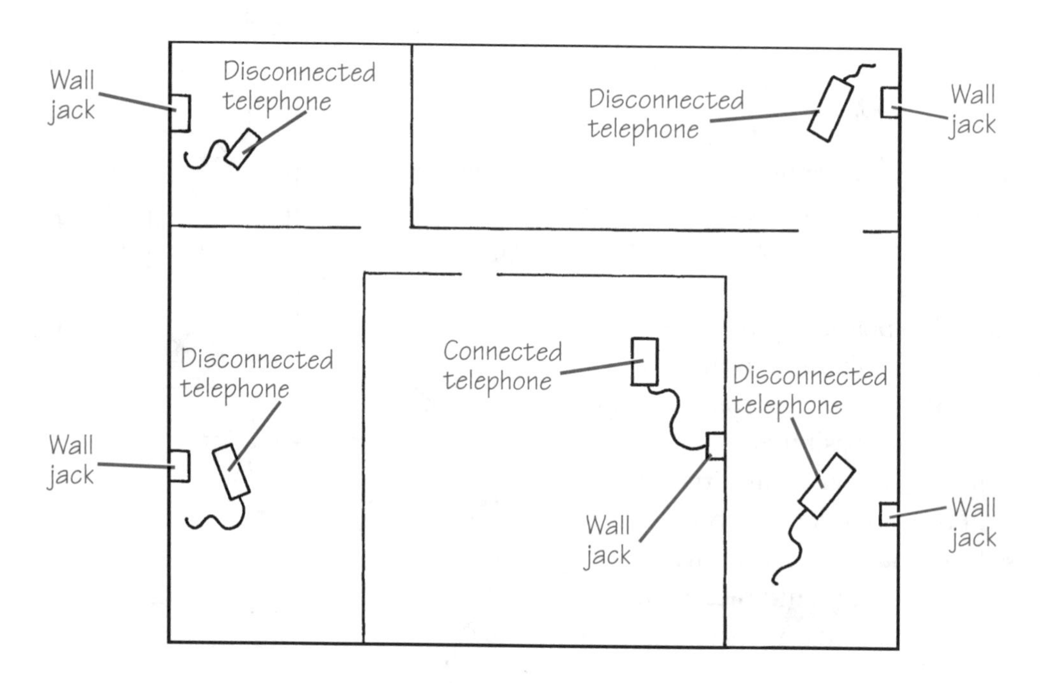

Some access boxes have a modular jack connected directly to the incoming lines. Sometimes this jack is beneath the box on the outside. Other times the jack is inside the box, and the box is locked (by the homeowner) to prevent unauthorized acess. If there is a modular box, you can use it as a quick way to find out if the problem is in the home or outside it. When a phone is plugged into this modular jack, none of the household wires or jacks are being used. If you plug in a phone that you know is working and nothing happens, it's almost certain that the problem is outside the home.

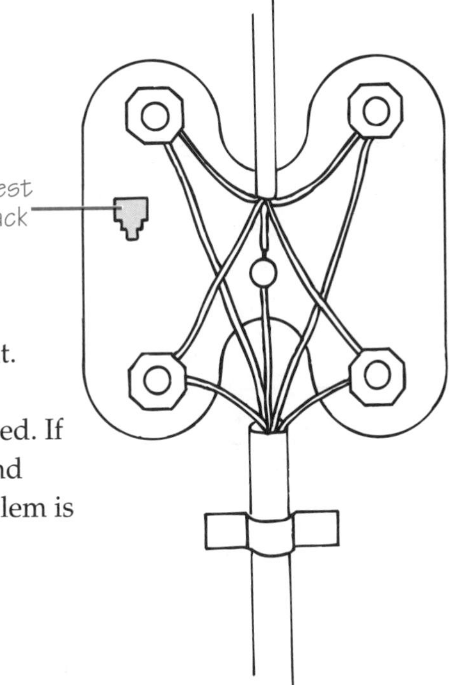

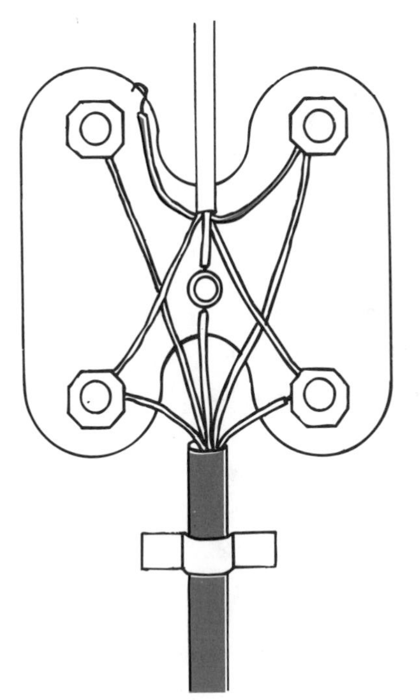

Step 4-2. Visually examining the wiring at the access box.
If even one wall jack is working, the problem is not in the access box. If none of the jacks work (and you're certain that the phone does work), the problem is either at the access box or with the incoming lines. Open the access box and visually examine the wires. Are they intact? Are the lugs tight?

Step 4-3. Looking for damage on exposed cables and cords.
Sometimes telephone wires are mounted to a wall. Because the cable is exposed, these wires are more prone to damage. If this cable has been cut or is otherwise damaged, you might have to replace it. (See chapter 5 for installation tips.)

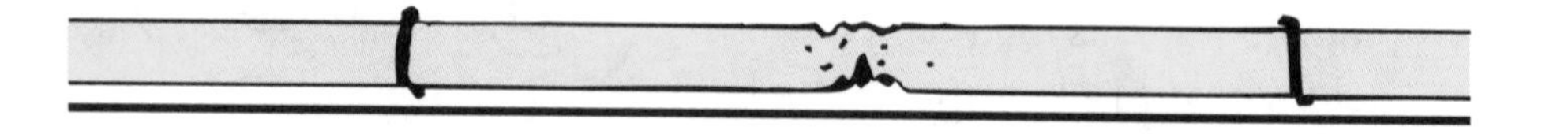

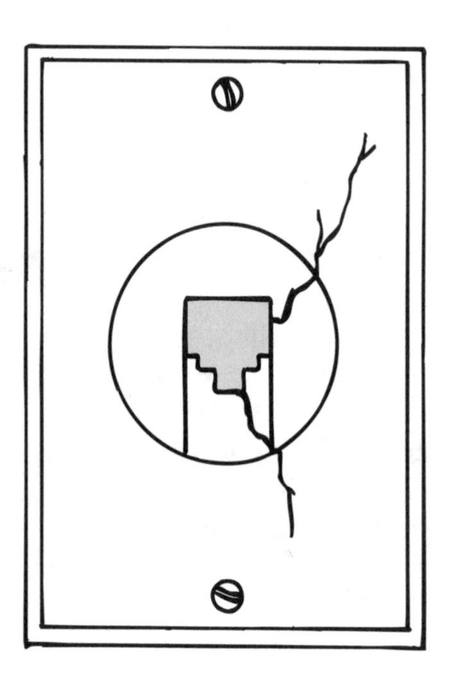

Step 4-4. Examining the wall jacks and plugs.
It's possible that some physical damage has been done to the jack or plug. Visually examine any plugs that do not seem to be functioning. The casing should be intact and the wires should appear clean and in good shape.

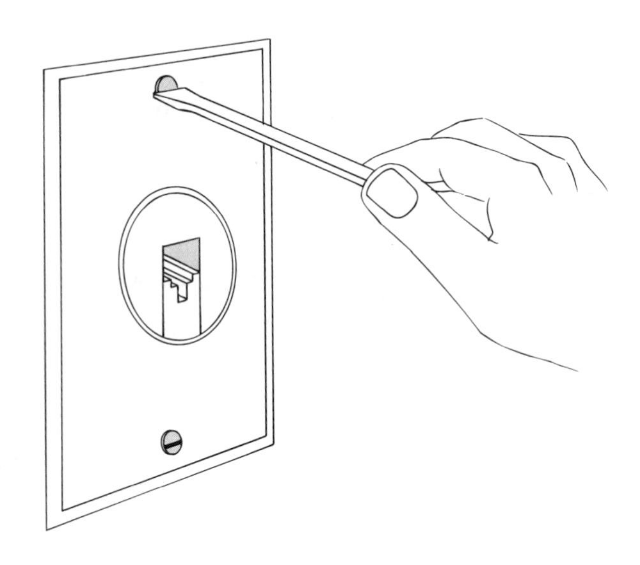

Step 4-5. Checking the inside of the jack.
The final visual check is with the inside of the jack. To do this, remove the cover plate and look inside. It should be clean with no signs of corrosion.

At least one pair of wires (usually the red and green wires) should be firmly connected to the jack lugs. If the second wire pair isn't attached, you can go ahead and connect them, but this usually will not be the problem. It's much more important that the correct wires are attached. (If the jack has ever worked, and no one has made any changes, this won't be the problem.)

The final visual check is with the inside of the jack. To do this, remove the cover plate and look inside. It should be clean with no signs of corrosion. At least one pair of wires (usually the red and green wires) should be firmly connected to the jack lugs. If the second wire pair isn't attached, you can go ahead and connect them, but this usually will not be the problem. It's much more important that the correct wires are attached. (If the jack has ever worked, and no one has made any changes, this won't be the problem.)

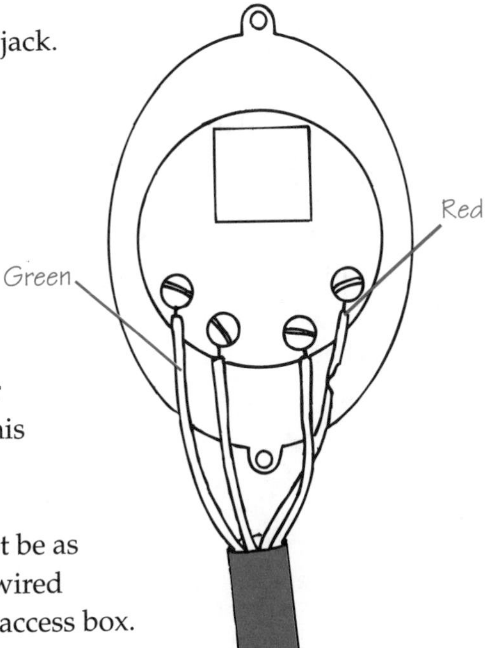

Note that the wiring order may or may not be as shown. You can find out how your system is wired (which color to which lug) by checking at the access box.

Step 4-6. Testing the telephone cords.
If you suspect that the phone cord is faulty, the easiest way to test it is by using a different cord that you know is good. If you don't have another cord, you can still test the suspected cord using a VOM.

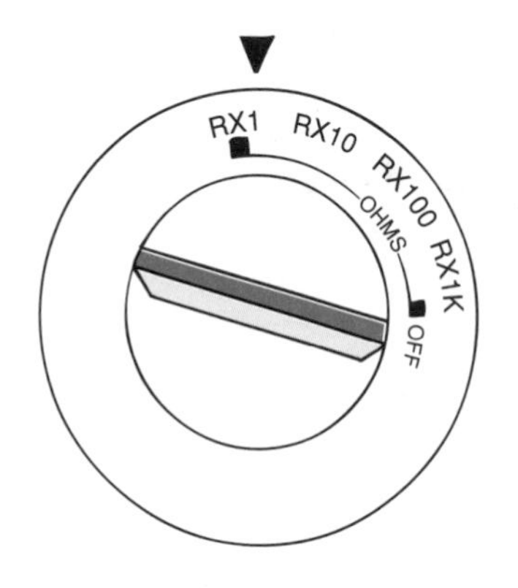

Set your VOM to read resistance (ohms). The scale used isn't important.

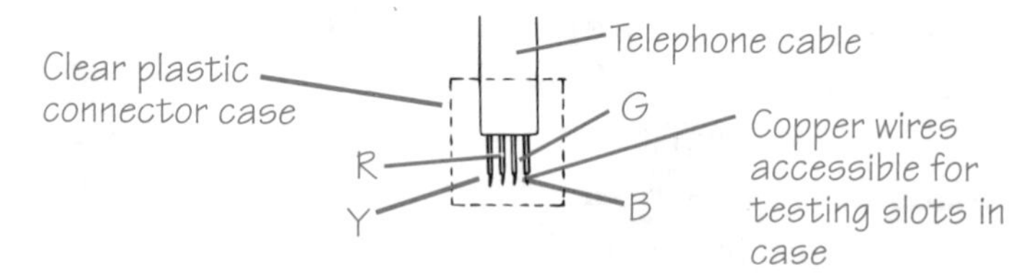

Disconnect the cord. Touch a pin on one side of the cord with one of the probes and the same pin on the opposite side of the cord with the other probe. You should get a reading of 0 ohms (the needle will swing all the way). This means that the wire is intact. If the reading is infinite or very high, the wire is bad and the cord must be replaced.

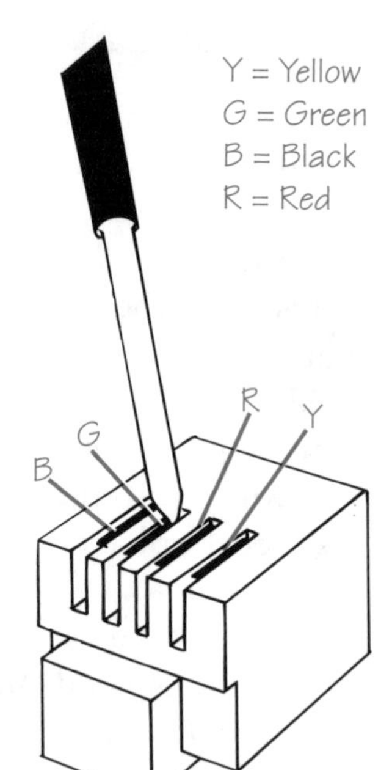

Both probes touching G.

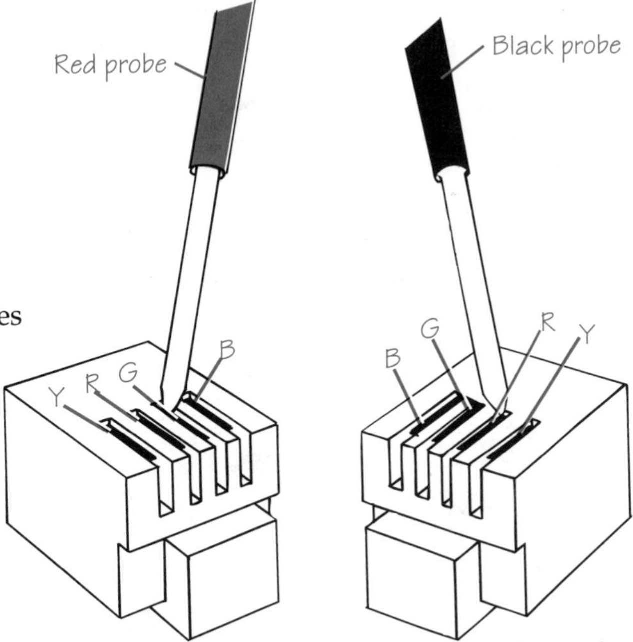

To test the cord for shorts (wires touching inside), touch a pin with one of the VOM probes and touch each of the other pins in turn. The reading you get should be infinite (no needle movement). This shows that the wires inside are not touching.

One probe stays on G while the other probe touches the other pins in sequence.

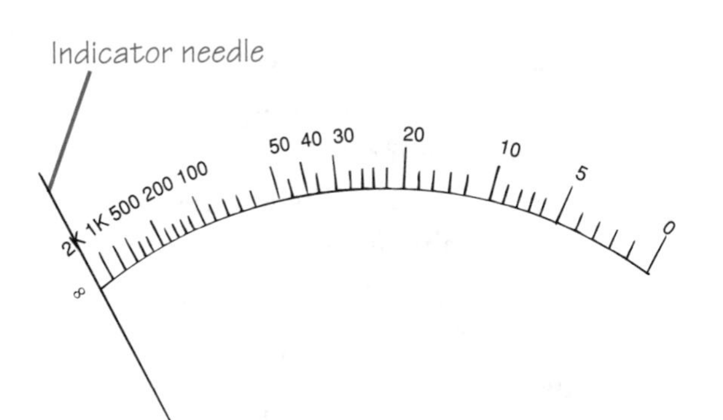

The reading is low, the wires are shorted and the cable must be replaced.

Repeat this test for each of the other pins. That is, move the first probe to the next pin and use the second probe to touch each of the other pins.

Step 4-7. Testing the jack and wiring.
To test the wiring inside the walls, or any long run of wire, you will need some way of putting a signal into the wire at one end and looking for it at the other. Appendix B shows you a very simple way to build and use a "signal generator."

Inside the house, begin by removing the cover plate of the suspected outlet.

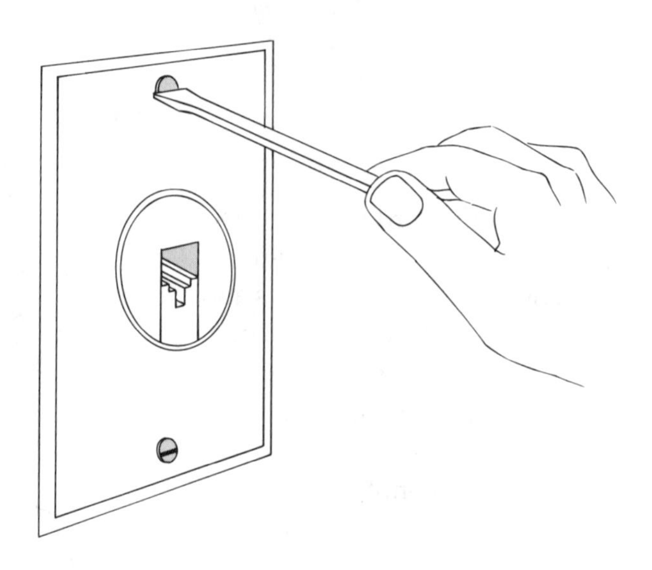

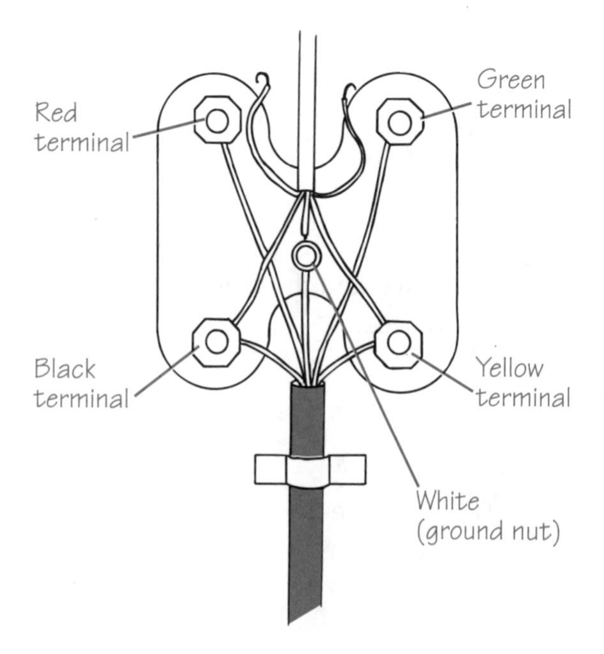

Outside the house, open the lid of the access box, loosen the outer lugs and remove the wire pair to be tested. Pull these two wires aside. You'll only be testing one wire pair at a time, but you can test all the wires, in any pairing, by repeating the test.

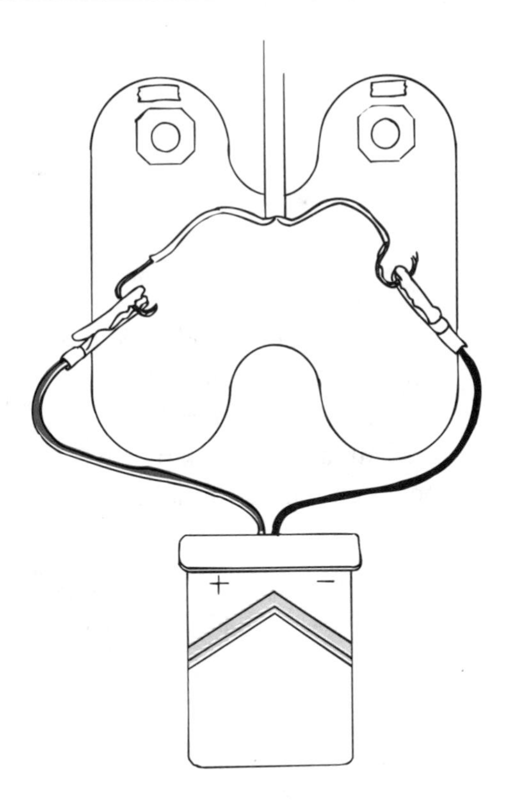

Attach the clips of the "generator" to the wire pair being tested. (It's a good idea to verify that the battery being used is good.) Note the polarity. In this example the positive side is the red wire and the negative side is the green wire.

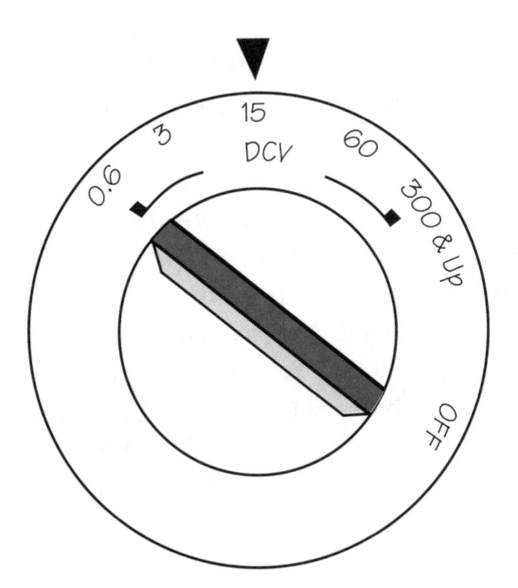

Back inside, set the VOM to read dc volts in the proper range. In this case, the signal is 9 volts dc, and the nearest setting for the meter is 12 Vdc.

Touch the VOM probes to the appropriate terminals (in this example, red probe to the red wire terminal and black probe to the green wire terminal).

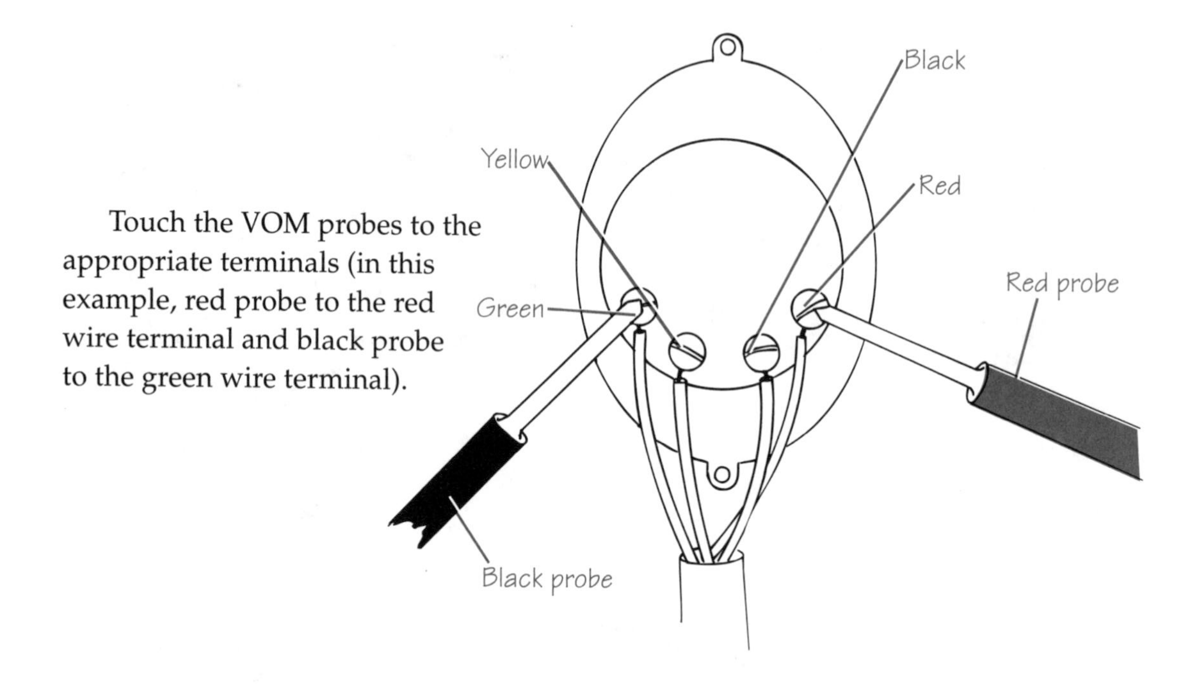

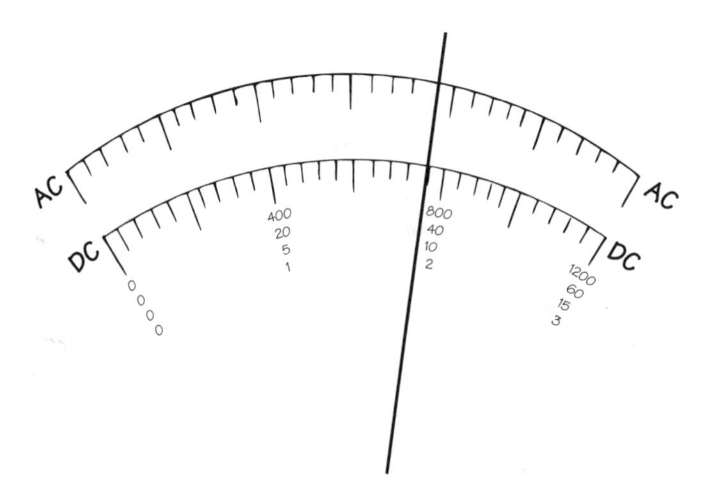

The reading you get should be 9 volts dc or very near to it. If there is no reading, or if the reading is very low, first verify that the battery is good. (This can be done by probing the battery itself directly with the VOM.) Assuming that the battery is good, an incorrect reading means that the wiring is bad and must be repaired or replaced.

If all four wires are connected at the access box and at the jack plate, move the black probe to the other terminals, each in turn. You should get no reading on the meter. If you do, the wires are shorted and must be repaired or replaced.

These tests can be repeated at each jack. You can also repeat them using other wire pairs. To do this latter, go to the access box and reconnect the "generator" to the other wire pair you wish to test.

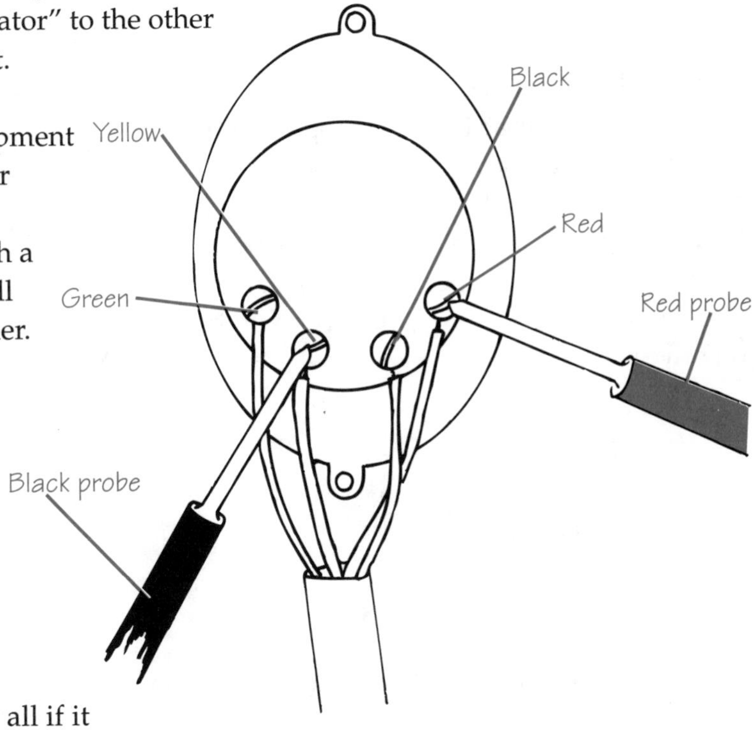

Some telephone equipment uses a power source other than the phone lines themselves. A phone with a memory and autodial will probably use a transformer. This plugs into the wall and converts the 120 Vac into the needed voltage. Other equipment (answering machines, cordless phones, etc.) can plug directly into a wall outlet.

In some cases, the device won't function at all if it isn't getting power.

Step 4-8. Checking the outlet.
If there is no power to the outlet, nothing plugged into it will work. An easy way to test the outlet is to plug something else into it, such as a lamp.

For a more accurate reading, you can test the outlet with a VOM (see appendix A). Set the meter to read volts ac in the appropriate range. (On this meter the setting is 150 Vac.) Insert the probes into the outlet slots. Because you are testing ac, probe orientation doesn't matter. Be sure to hold the probes by the insulated handles only. The reading you get should be on or very near 117 Vac.

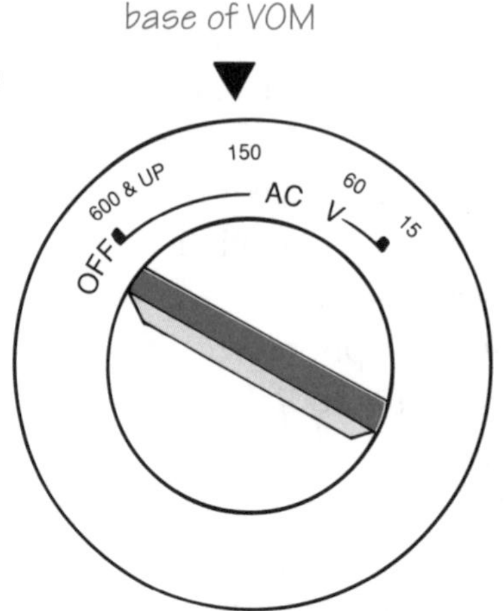

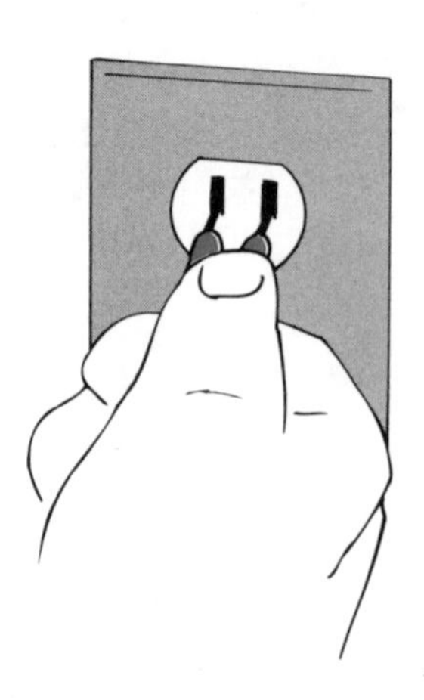

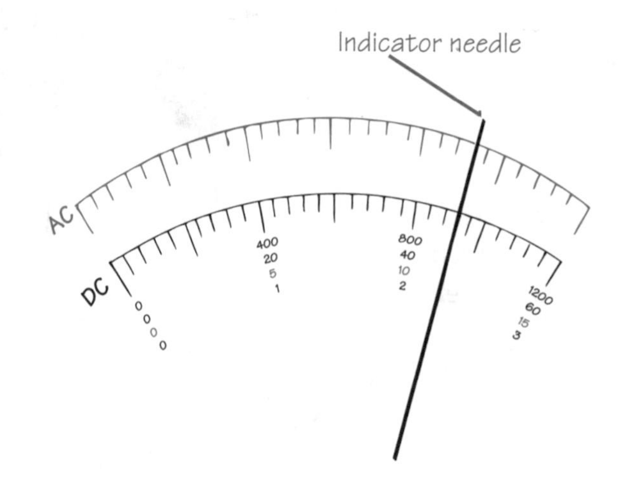

Step 4-9. Checking the wall switch.
If the outlet is dead, look to see if the outlet is operated by a wall switch. Some homes have switches so that plug-in lamps can be turned on from a switch.

Step 4-10. Checking the fuse or circuit breaker.

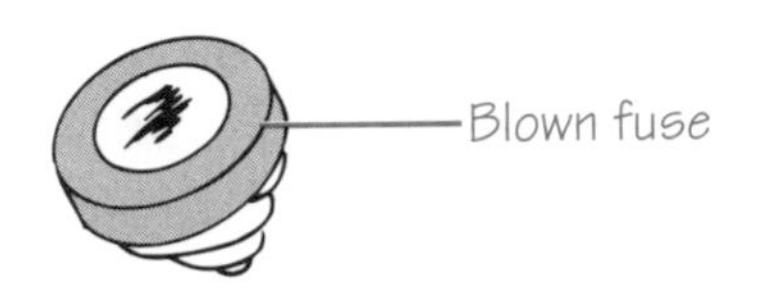

Power flow through the house is protected by either fuses or circuit breakers.

Usually you can see if the fuse wire is melted by looking through the "window" of the fuse. A popped circuit breaker might not be quite so easy to see.

Normally the lever will be slightly back, but sometimes so slight that it is difficult to see. When in doubt, flip the breaker completely off and back on again.

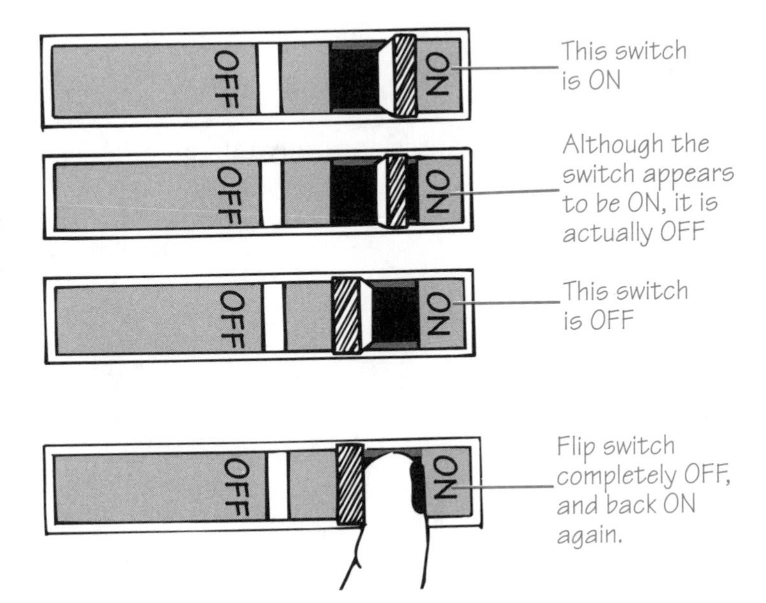

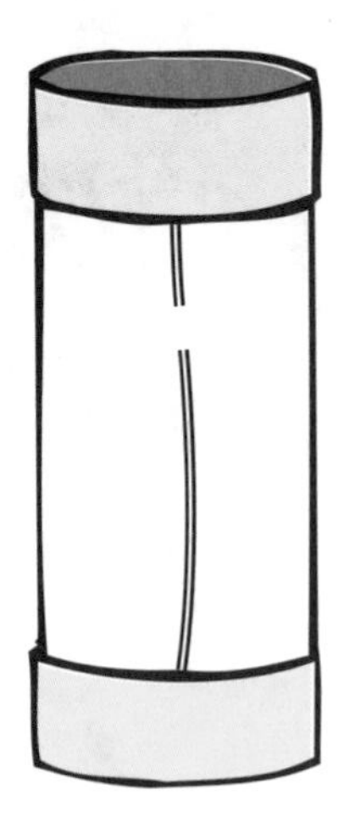

Step 4-11. Checking an internal fuse.
Some devices have an internal fuse. With telephone equipment this usually means taking the equipment apart. If the outlet is good but the device still won't "power up," this internal fuse might be blown. In many cases you will have to open the cabinet to get at the fuse. As with the household fuse, you can usually just look through the glass to see if the fuse wire inside is melted.

Step 4-12. Testing the fuse with a VOM.
Sometimes a fuse will look good when it is actually bad. If you suspect this, shut off all power and unplug the computer. Remove the fuse from its holder.

Set your VOM to read resistance; any range will do.

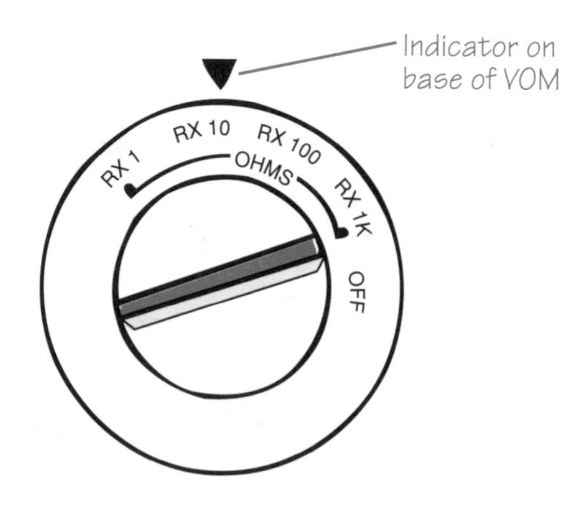

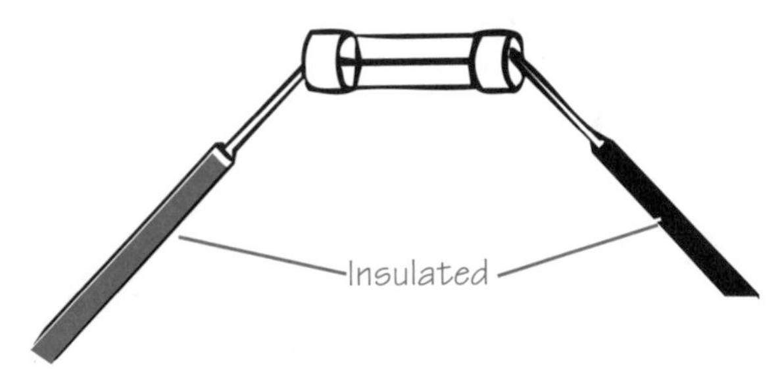

Touch the probes to each end of the fuse. You should get a reading of nearly zero ohms. If the fuse shows a high resistance, it has "blown" and must be replaced.

Step 4-13. Testing a transformer.
Many telephones and related devices are powered by a transformer, which drops the voltage from the wall outlet to the voltage level needed by the device.

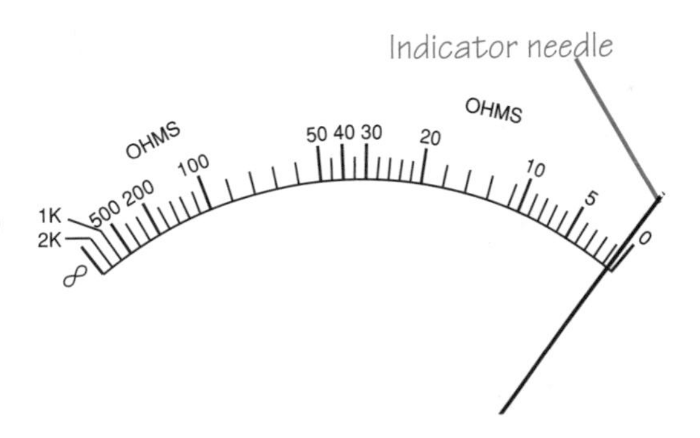

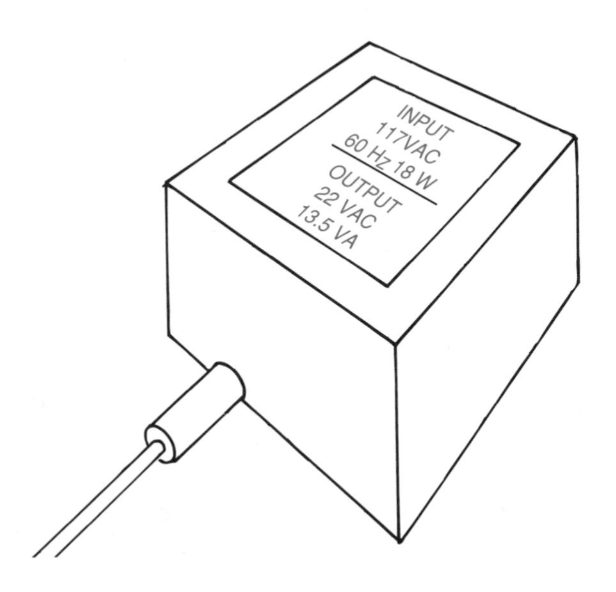

The transformer will almost always have a label on it that gives the specifications.

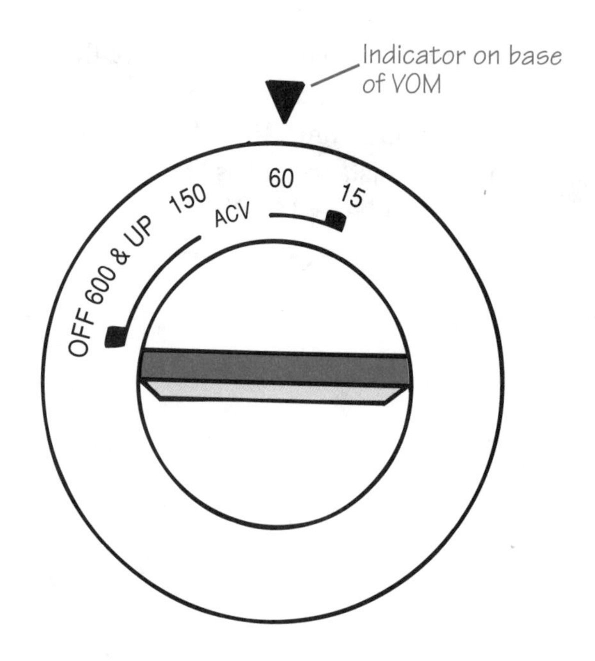

Set the VOM to read volts ac in a range above and closest to the transformer output. This will differ from one VOM to another. In this example, the closest range is 60 Vac.

With the transformer still plugged into the wall outlet, disconnect the transformer's connector from the device.

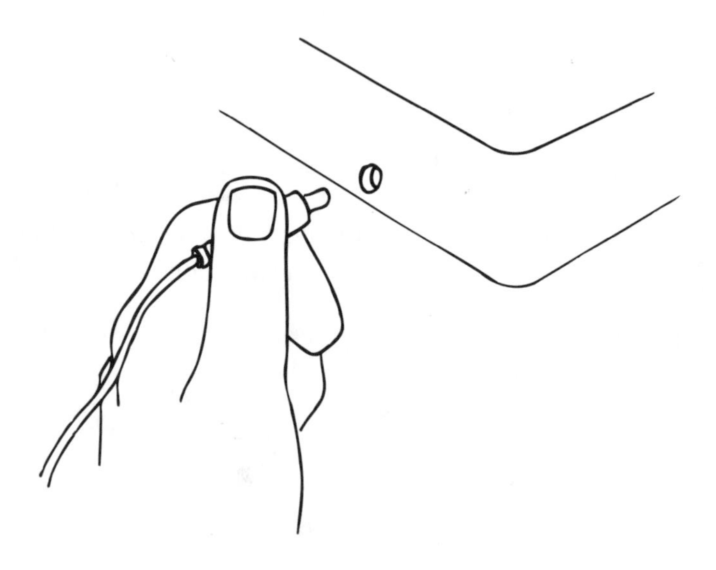

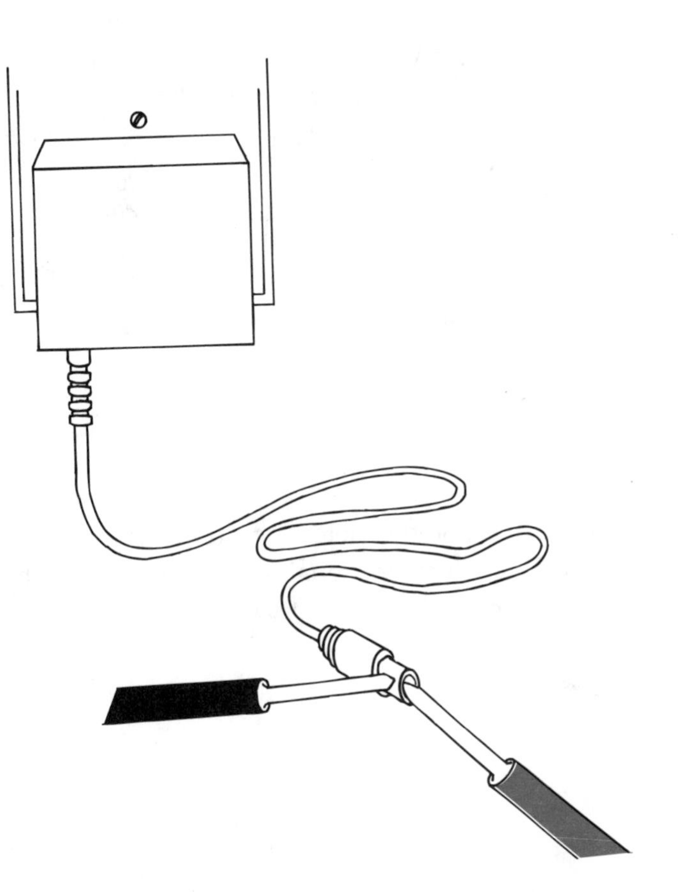

Touch one of the VOM probes to the inside of the plug and the other probe to the outer metal shell. Because this is ac, polarity of the probes doesn't matter.

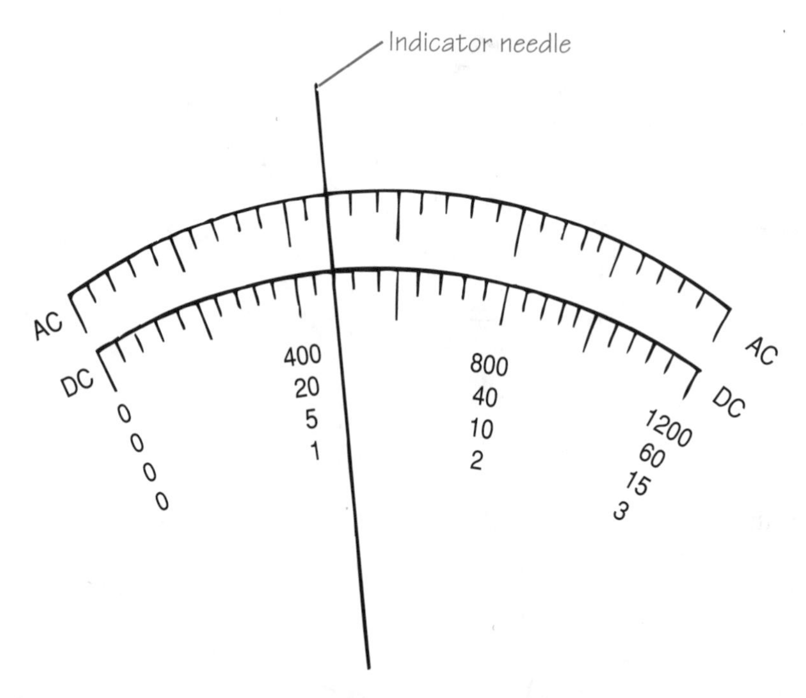

The reading on the meter should be very close to the rated value. With most transformers, repairs are impossible. A bad transformer will have to be replaced.

CHAPTER FIVE

Wiring

The wiring of a household telephone system isn't complicated. In most cases a cable holding four insulated wires will come to each jack. Even if all four of these wires are connected, only two of them are used for a particular phone line.

It's important to note that your particular system might be different. For example, the cable coming to the outlet might have six wires or even eight wires. The colors might be different, or they might be used in a different way. Don't let this confuse you. However it's done, the job of the wiring is to connect two wires from the phone company to the telephone. *This* is your primary concern.

If the cable has four wires or 16 in it, only two of them are really being used. The others are there for additional phone lines (different phone numbers) and/or as back-up wires. If the insulation is colored yellow, green, purple, or whatever, the wire inside is still doing the same thing.

Usually the pair that is used is consists of a red and green wire, with the red one attached to the upper left and the green to the upper right. If the yellow and black wires are connected, most often the yellow is connected in the lower left and the black in the lower right.

Occasionally, especially in older systems, there will be a ground wire. This wire is often white. In this case, the cable may contain only three wires. The next most common configuration uses the yellow wire. Either way the usual lug will be lower left. That is it will be *if* the household system uses a ground at all. Many systems used today are grounded only at the access box (and at various places throughout the telephone company's system). This is why you may hear the access box also called the "house protector."

Step 5-1. Determining the incoming lines.

The telephone company's line can come in overhead (aerial) or it can be buried. Except for basic troubleshooting, it doesn't matter either way. Your own troubleshooting of any lines owned by the phone company is limited to looking to be sure that an aerial drop hasn't been damaged, or that a buried line hasn't been cut. Beyond this, you need not be concerned with this line.

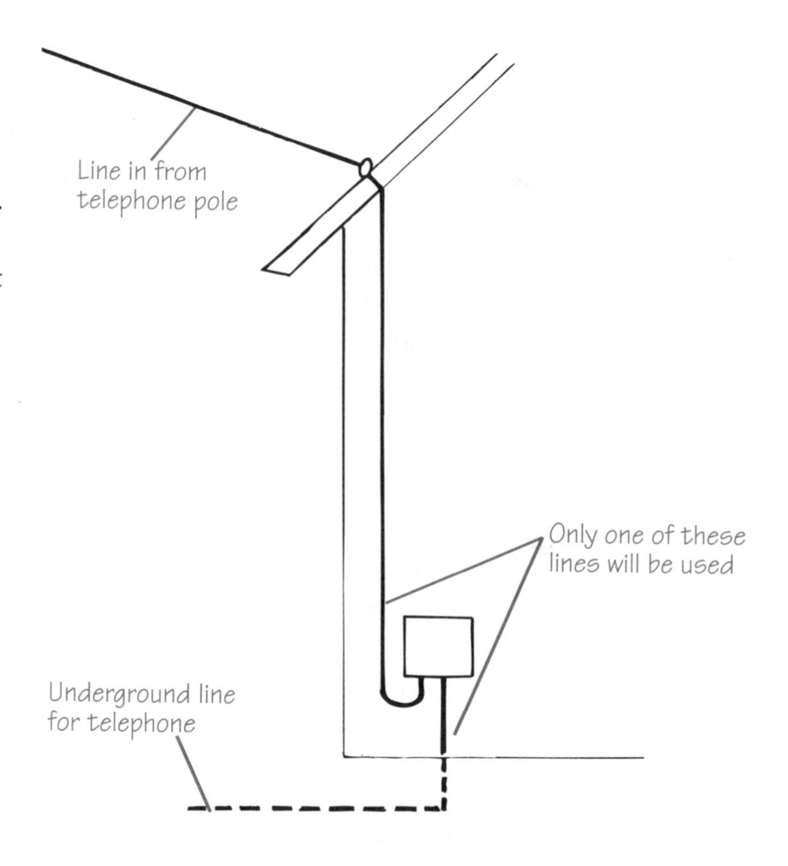

You can often get a clue as to how your system is wired by looking inside the access box. Even though there are several different kinds, the basic function inside the access box is pretty much the same. The telephone company's drop line enters the box and is attached to the lugs inside. The line that goes into the home is attached to the same lug shaft. Almost invariably, the phone company's line will be farthest back and it will be larger.

By merely looking you'll be able to tell if two, three, or four wires are being used, and what the color coding is.

It's important to note any grounding that is used. Sometimes this goes through fuses. Other times there will simply be a grounding bar or wire. Do not disturb this part of the system unless it has been damaged.

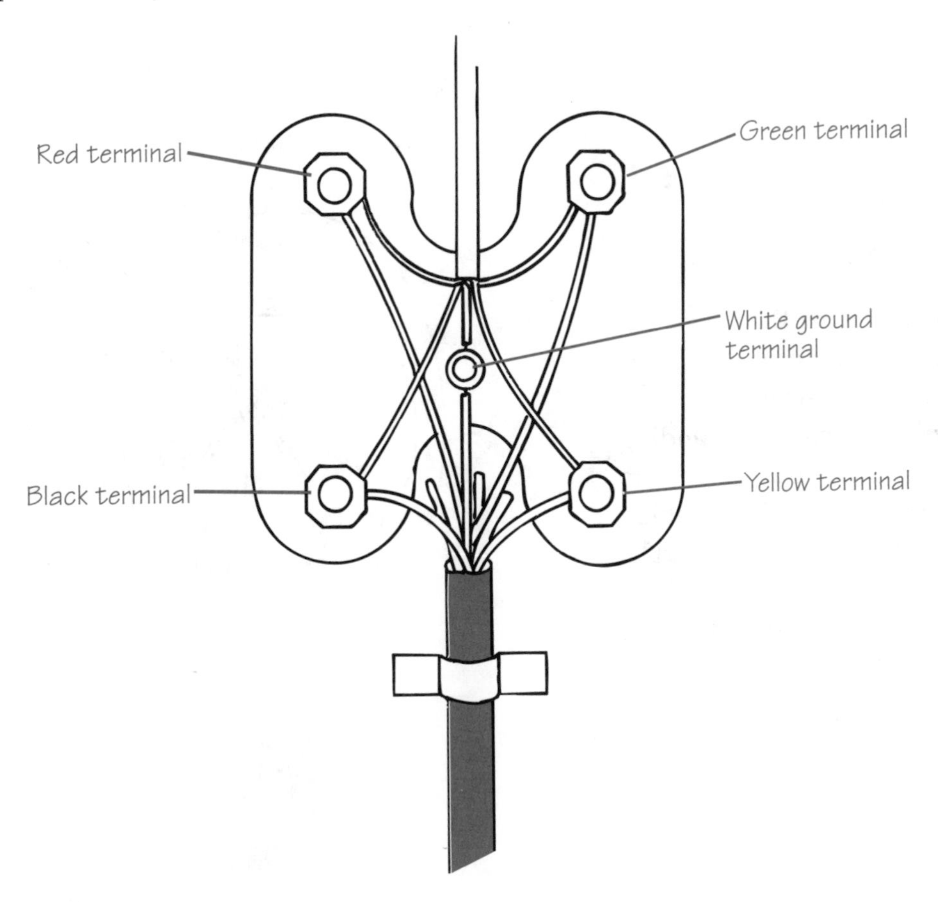

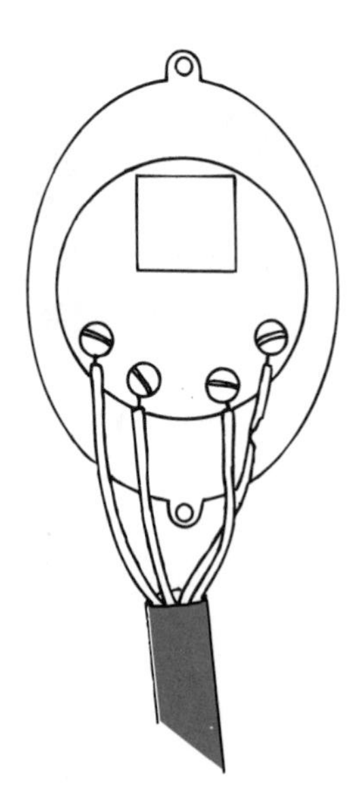

You can verify this by looking inside a jack in the home. Unless someone has wired the home strangely, the wall jack wiring scheme will be the same as at the access box.

Step 5-2. Finding the distribution pattern. "Finding the distribution pattern" is a technical-sounding way to say that you will need to know how the telephone cable moves through the walls and to the jacks. You can often tell by looking inside a wall jack.

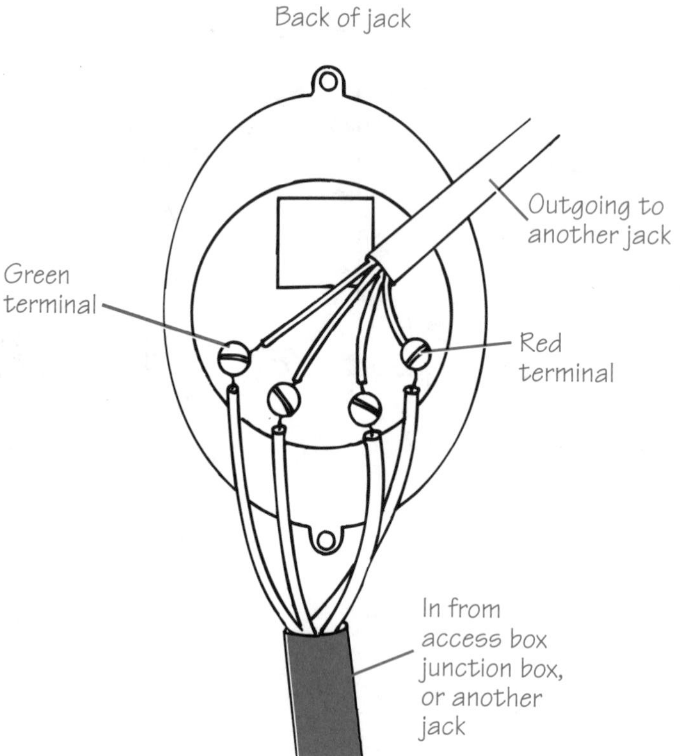

The series pattern is common because it is less expensive to install. The cable comes from the access box and goes to each outlet in turn. You can identify this pattern by two distinctive cables: one coming in, and one leading to the next outlet.

Series wiring

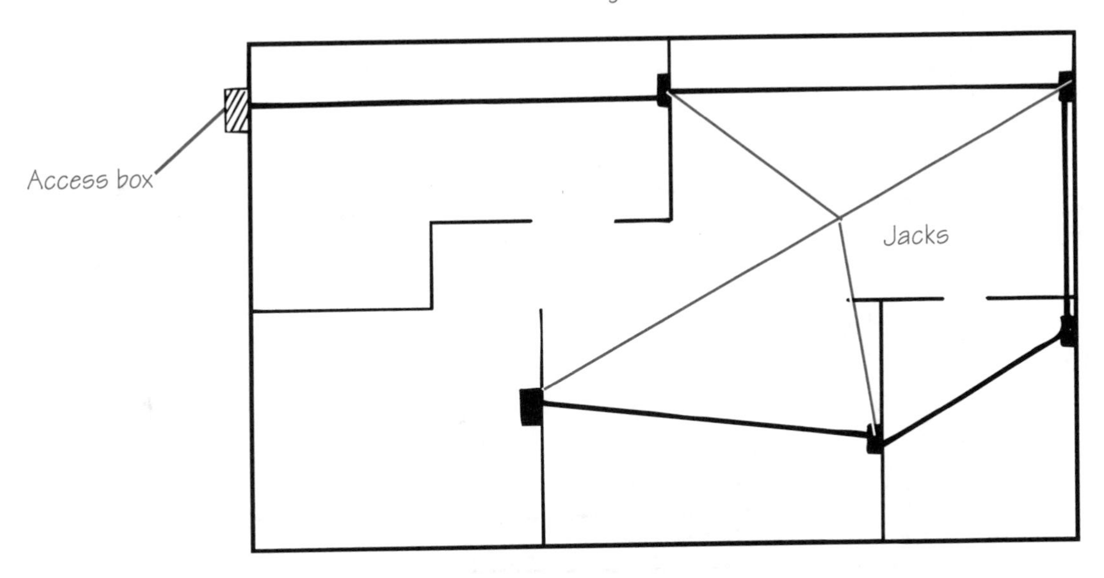

"Home-run" wiring uses a separate cable for each phone jack in the home. Sometimes this is done by connecting each of those cables to the lugs in the access box.

Parallel wiring

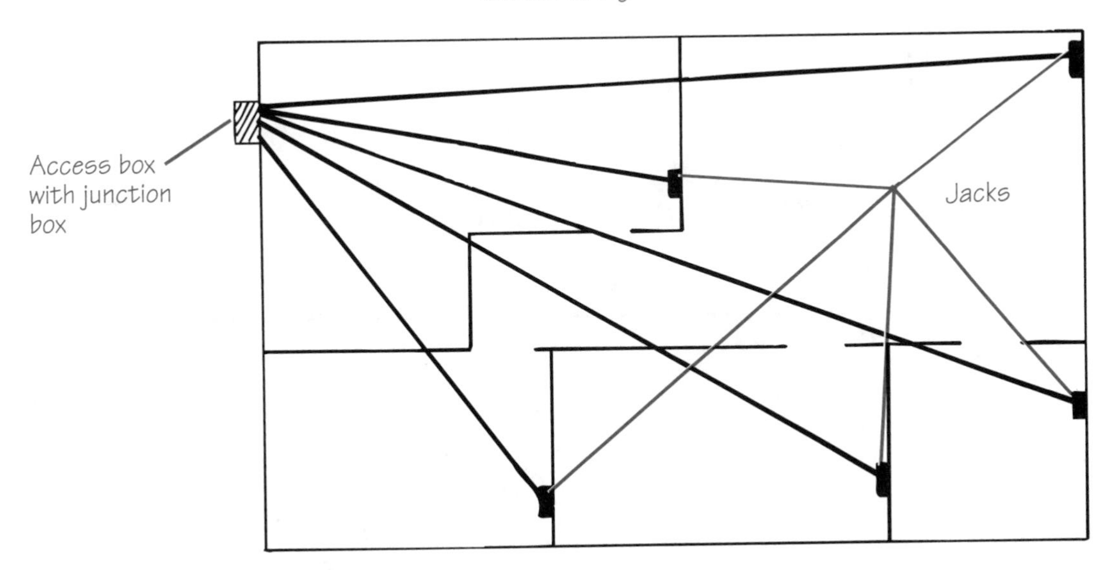

Other times a single cable will leave the access box and be routed to a main junction box and be "split" there. The location of this box can vary.

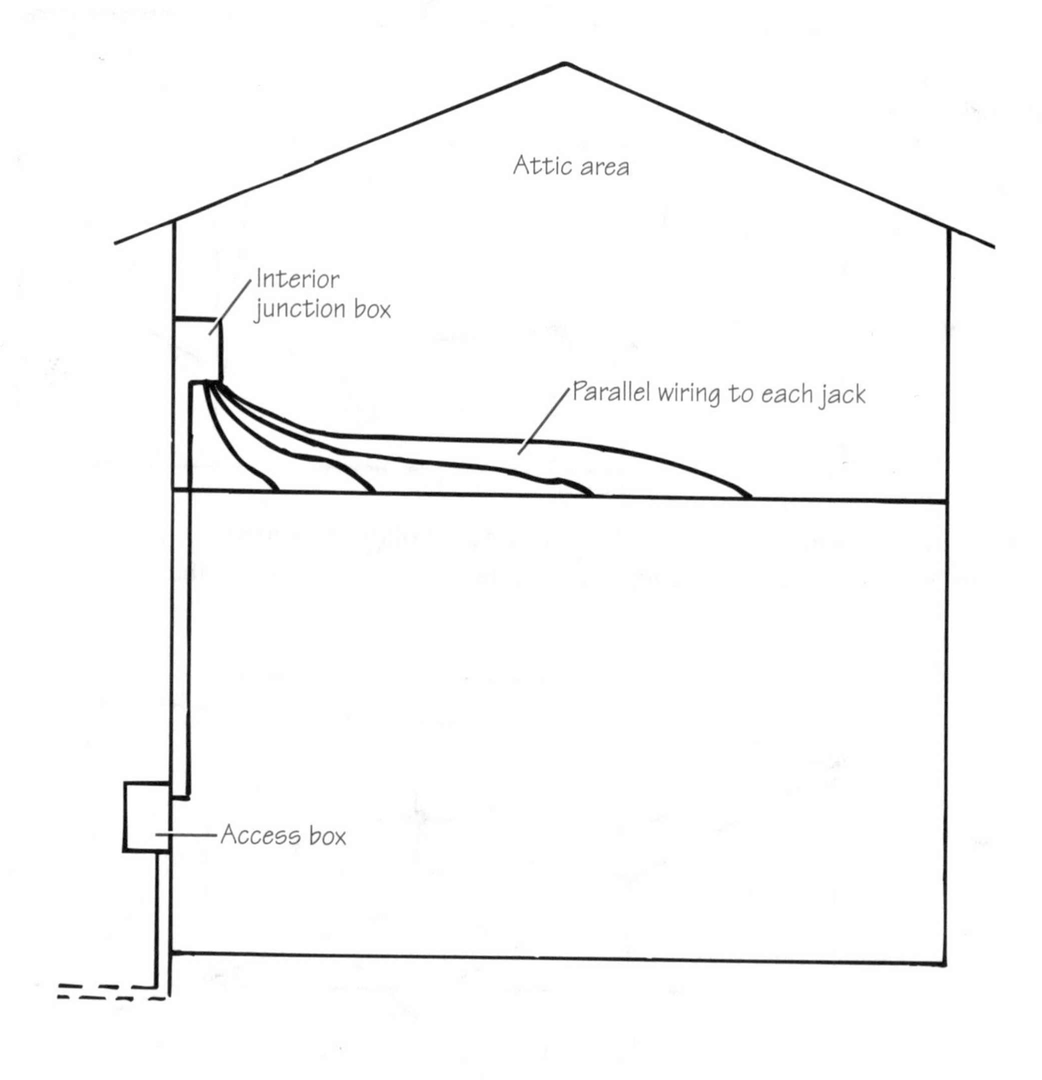

Step 5-3. Finding or routing the wires.

If a damaged telephone cable has to be repaired or replaced, or if you wish to install a new jack, you need some knowledge of how the cables go from the access box to the individual jacks

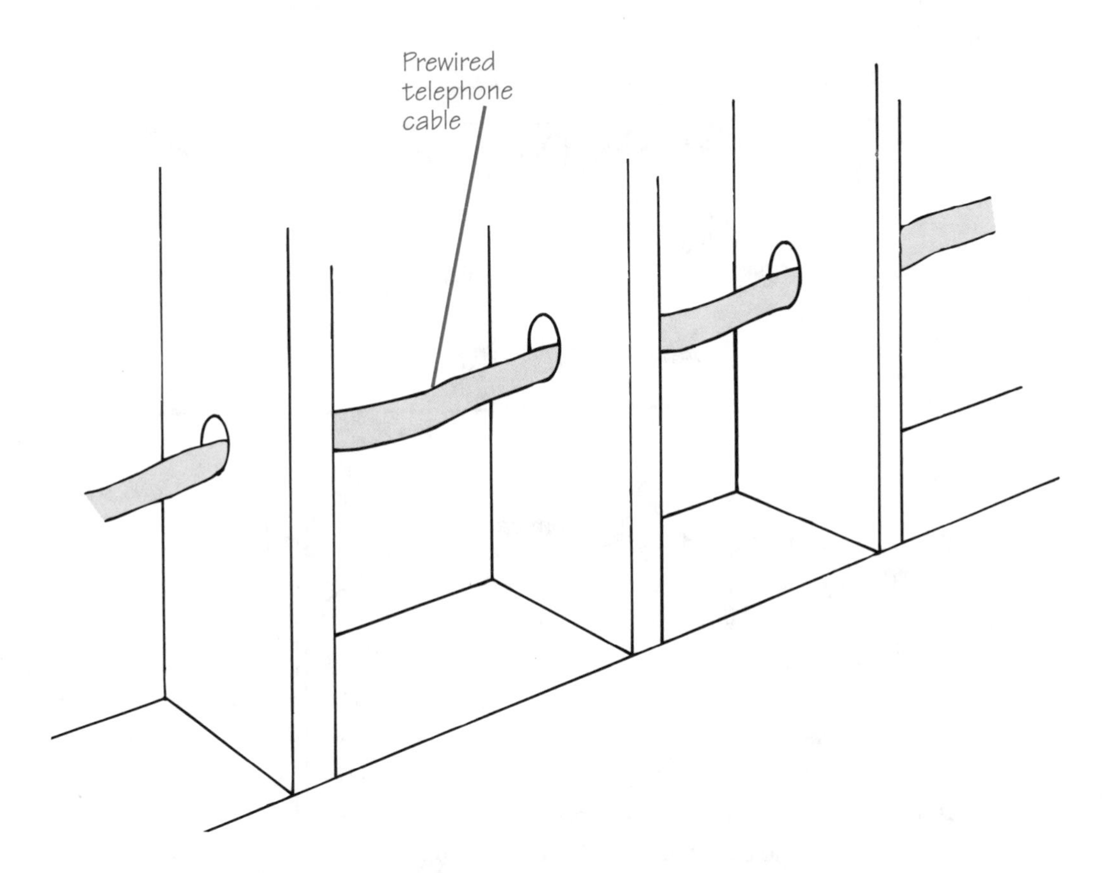

The routing of the wires depends largely on when the wires were installed. If put in place during construction, it is called "prewire," and the cables will be inside the walls, often passing through small holes drilled through the supporting studs. Repair is usually impossible. You will have to use one of the following methods to install a new cable as a replacement or to install a new jack.

Sometimes the wiring will be above or below the home. Usually they will be routed to the wall where the jack is located and will then pass inside the wall space. (The details of installing the jack are given in the next chapter.) Note that the cable should be well secured, usually with staples. If someone else has done this and the cable is causing problems, it could be that the staples were put in too tight or have nicked the wires inside the cable. When you are installing a cable, be sure to avoid these problems.

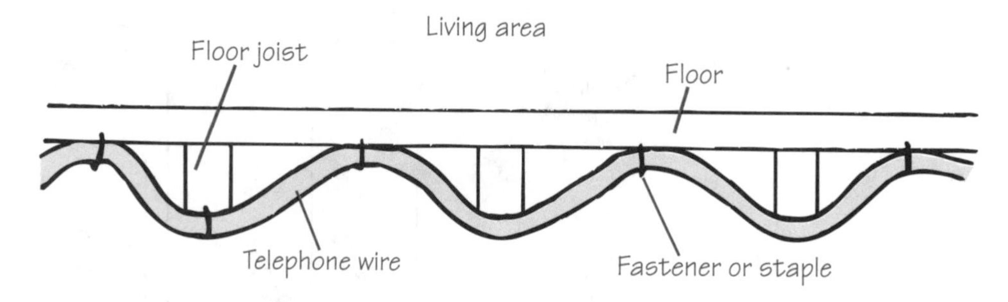

One of the easiest methods is to route the cable around the outside of the house until it comes to the exterior wall of the room where the jack is located or will be installed. A hole is drilled through the exterior wall at this point. After the cable is inserted, the hole around the cable is sealed.

This is an easy method, but it's also the least sufficient. Having the cable exposed this way can lead to damage. When a problem occurs, examine the cables carefully. If you're installing a cable this way, try to attach it up under the eaves rather than close to the ground.

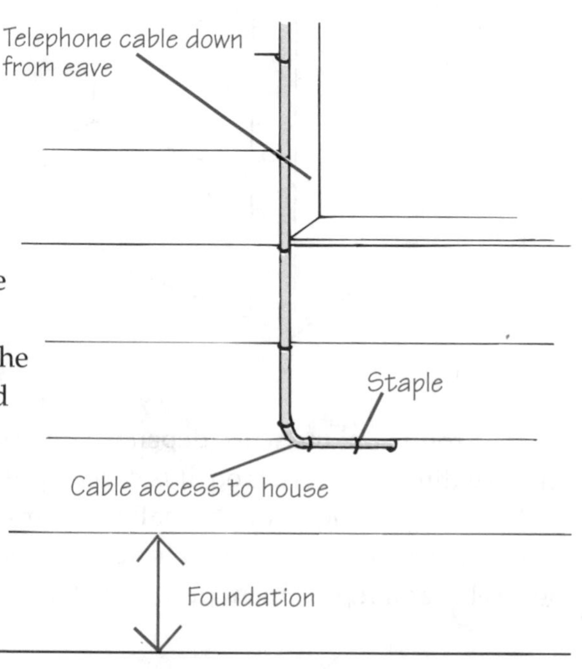

Step 5-4. Installing the cable.
When you are replacing a damaged cable, the new cable should follow the route of the old one. (This probably won't be true if you are replacing a prewired cable that was installed inside the wall during the home's construction.) Routing a cable for a new phone jack will be determined by where the jack is to be located, and by the circumstances. In most cases, it is preferable to run the cable either through the attic or in the crawl space; use outside routing as a last resort.

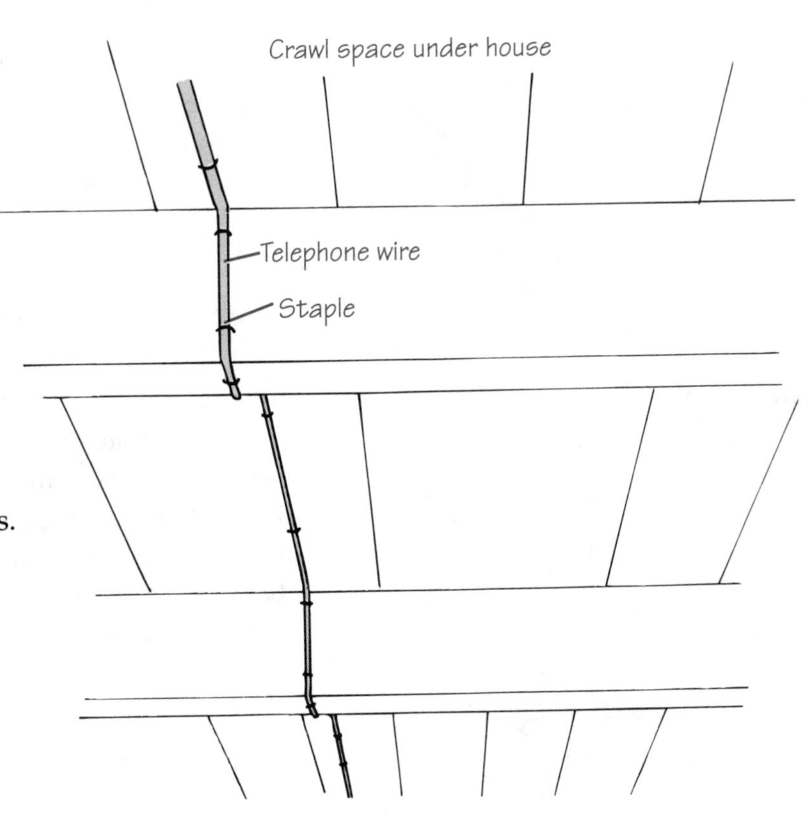

When running the cable in a crawl space, don't just let the cable lie on the ground. Attach it to overhead supports, such as the floor joists.

Stapling the wire isn't as critical in an attic but it is still a good idea. If possible, place the cable where it is out of the way, such as over to one side, or staple it to the rafters overhead.

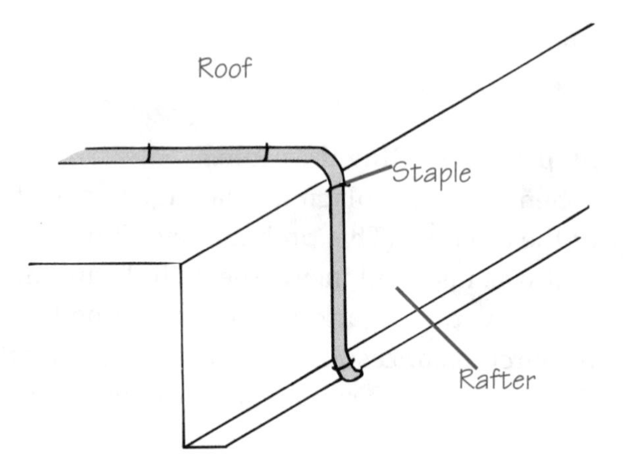

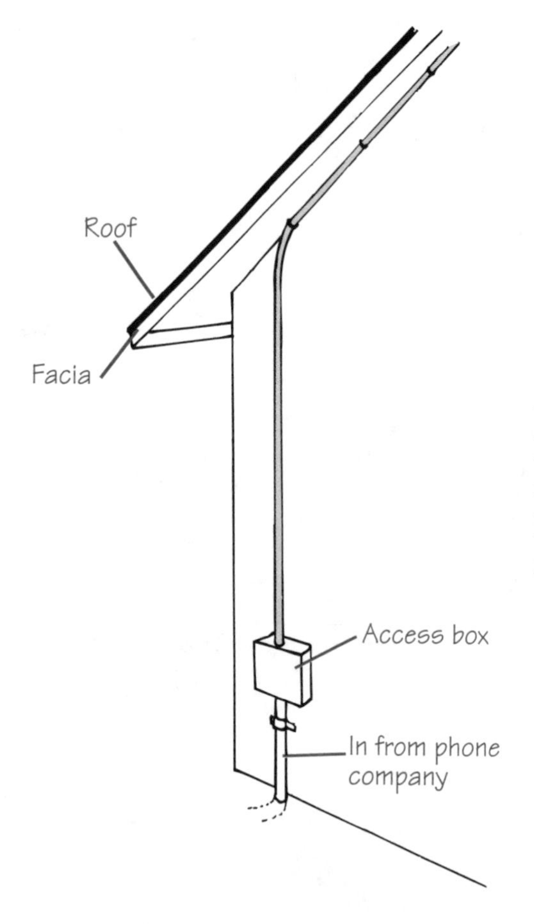

If the cable must be routed outside, it is usually better to run it up to the eaves. This keeps it off the ground and puts it where it is less likely to be damaged. Placing it at the bottom of the eaves and against the house makes the wiring less visible.

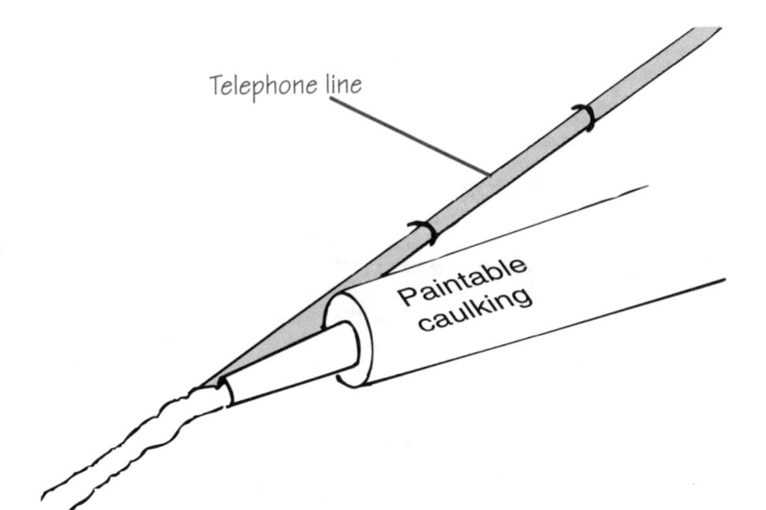

To make the installation more attractive, the wire can be covered with caulking. It is preferable to use the kind of caulking that can be painted. This way you can make the cable almost disappear.

When the cable must be run along an inside wall, an easy way to hide it is to tuck it beneath carpeting. The best way to do this, however, is troublesome. It involves lifting the carpet off the tack strip that holds it and then pushing the carpet back into place. This can cause the carpet to loosen if you don't re-install it correctly.

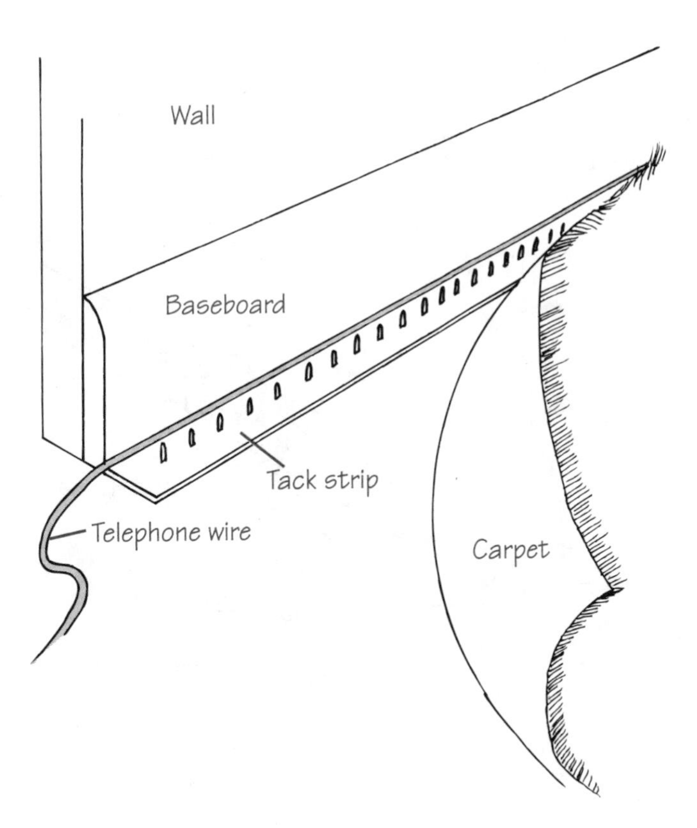

When there is carpet, the usual option is to attach the cable to the top of the baseboard. You can paint the cable to match the wall or baseboard to make it less visible.

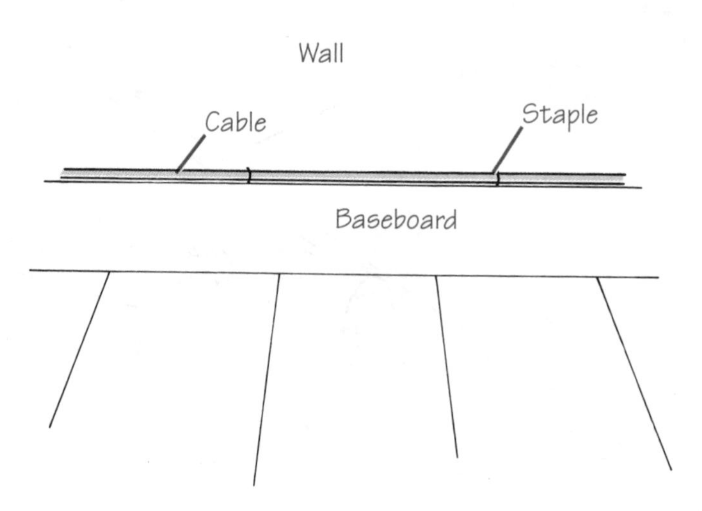

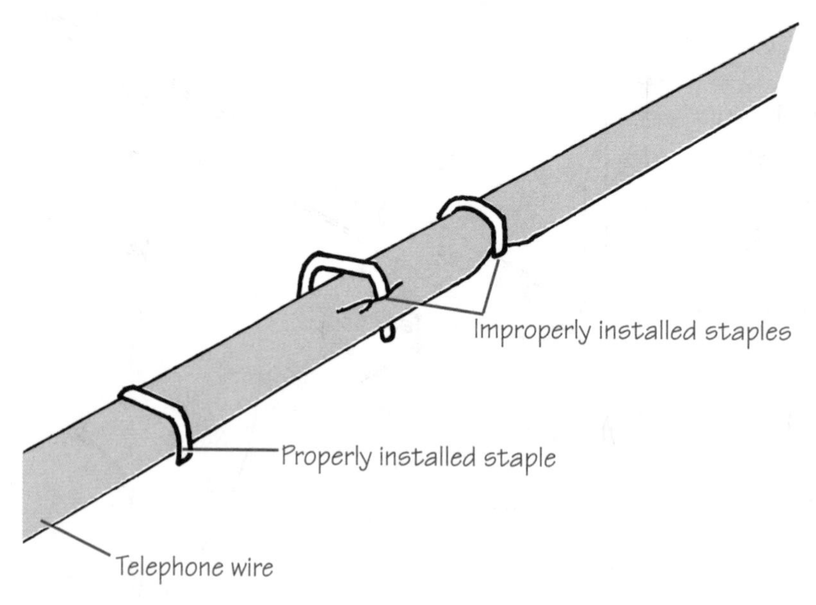

Stapling is a fast and easy way to secure the cable. The staple gun and staples used must be strong enough to penetrate the material to which the cable is being attached. Take care to get the staple all the way in, but don't make it too tight. Also make sure the staple doesn't cut or damage the cable.

If you don't have a staple gun, you can buy staples that are meant to be installed with a hammer. To save your fingers, a pair of needle-nose pliers can be used to hold the staple in place. Again, avoid cutting or damaging the cable.

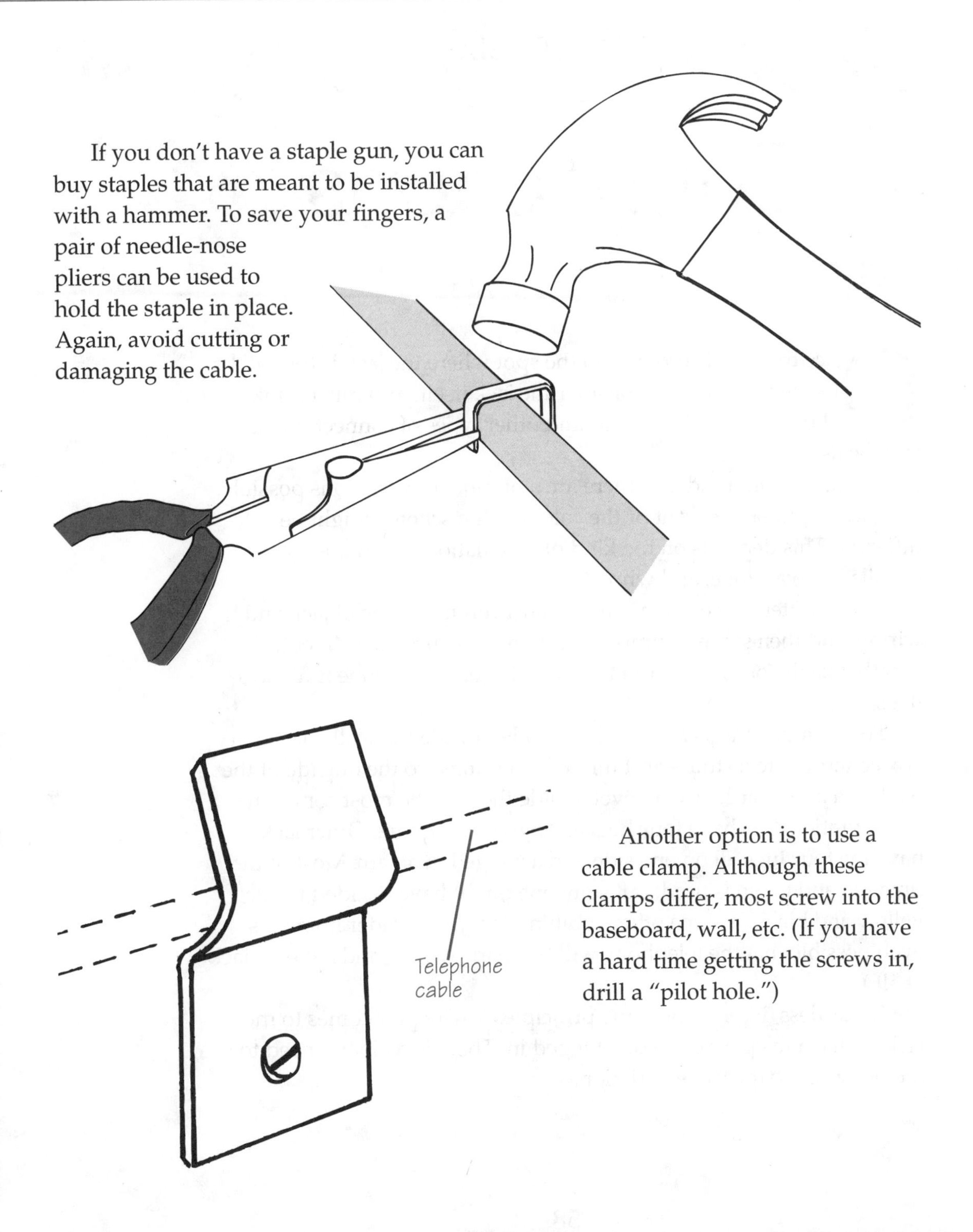

Another option is to use a cable clamp. Although these clamps differ, most screw into the baseboard, wall, etc. (If you have a hard time getting the screws in, drill a "pilot hole.")

CHAPTER SIX

Connections

Once the wire is brought to the spot where the jack is (or will be) located, it's time to consider the jack itself. You must think about the way it is wired and other types of connectors that may be used.

Again keep in mind that there are a number of variations possible. The jack might be different or the color coding scheme might be different. This depends on the kind of installation, when the installation was done, and who did it.

This chapter concentrates on the most common type of jack and wiring, and then some common variations. If yours doesn't look exactly like the ones here, don't worry. The job being done is always the same.

For example, the jack might be installed inside the wall and use a box connected to a stud—or it might be mounted to the outside of the wall, to a baseboard, or whatever. Inside the jack the most common configuration is to have four lugs arranged in a square. Other jacks have the four lugs all on one side and arranged in an arc. Most of the time red and green is used, but someone might have decided to use yellow and black or some other combination. (In an older home it's even possible that the telephone will be "hard-wired" and have no jack at all.)

Regardless, it's still the same principle. A wire pair comes to the jack so that a telephone can be plugged in. There is really no need to memorize every possible variation.

Modular and four-prong jacks

Quite long ago telephones were hard-wired. The cord from the phone went into a connecting box where the wires were attached. If the phone had to be replaced, the connecting box was opened and the wires taken out from under the lugs.

To make servicing easier, a four-prong plug and jack was invented. The plug and jack are *keyed*. That is, there is a pattern which makes it impossible to plug in the phone incorrectly.

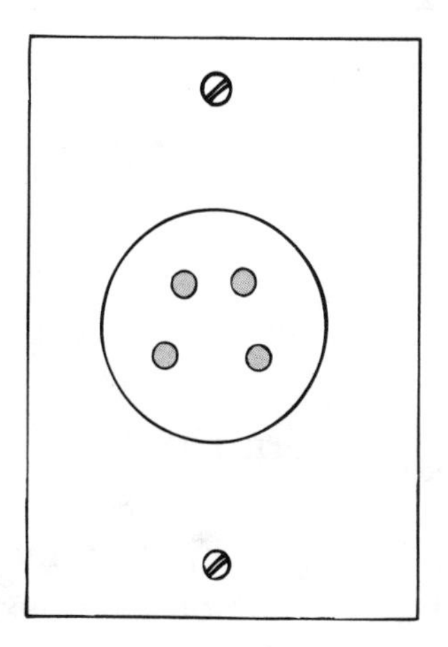

Flush-mounted four-prong plug

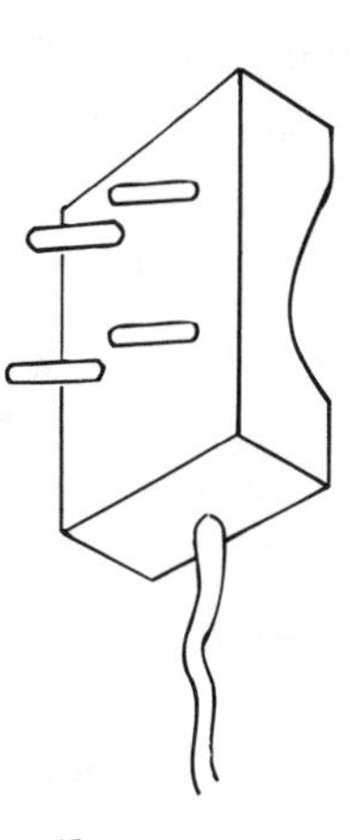

Four-prong jack

These days almost every telephone and telephone-related device (answering machine, computer modem, etc.) uses a modular plug. If you buy a modular plug, you'll have to have a modular jack or at least an adaptor. (It is better to replace the jack. This is covered later in this chapter.)

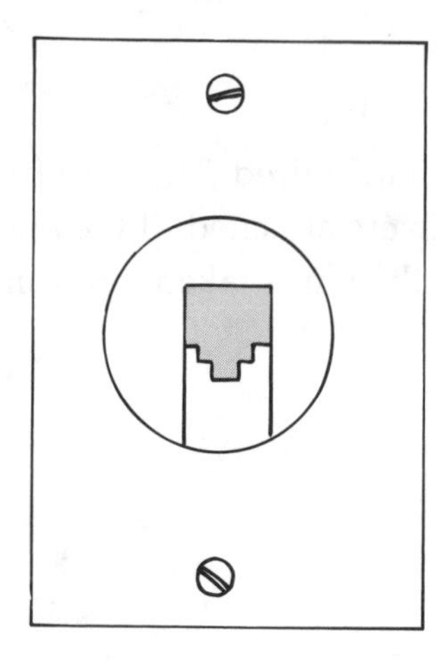

Modular plug

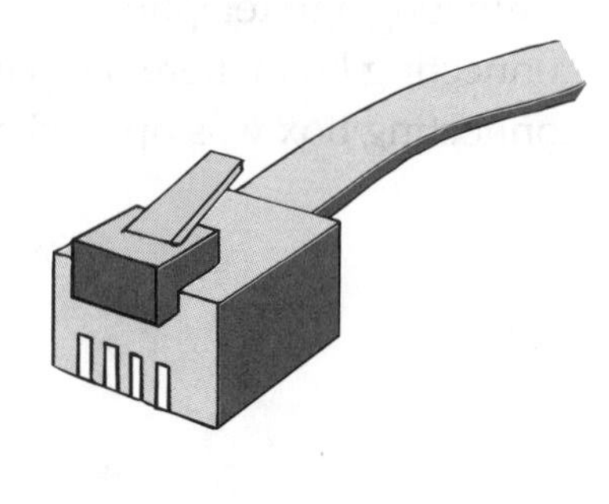

Modular jack

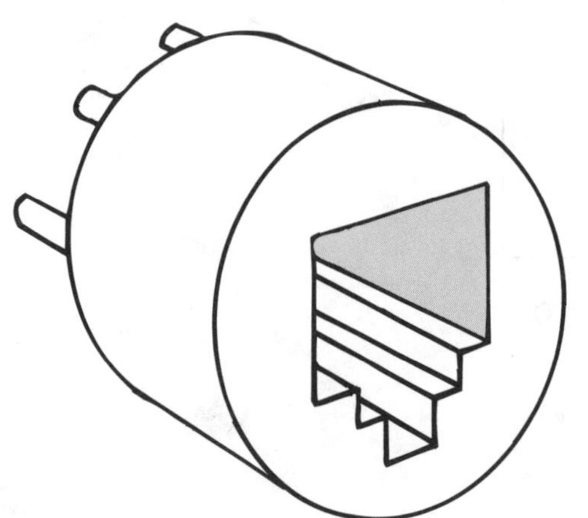

Four-prong to modular adaptor

The easiest way to adapt an older four-prong jack for modular use is to use an adaptor. This inexpensive device has four prongs on the plug side and a modular jack on the other.

Two common jack plate types

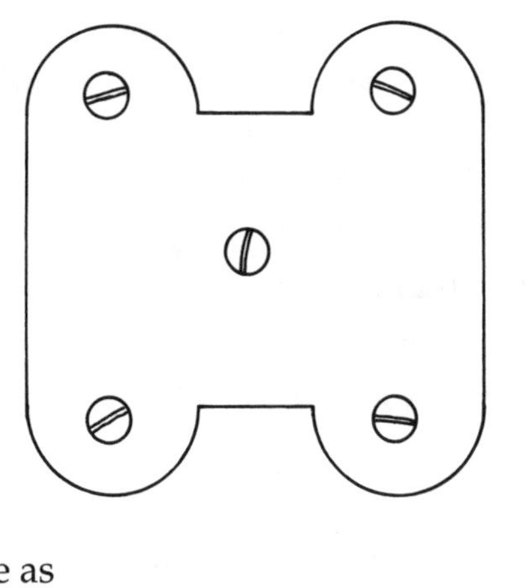

The two most common jack plate configurations are square and arced. Either may appear as shown or in some similar fashion. The same is true of how the wires are connected. You will have to look at your own configuration to see how it is connected. Most often it will be as shown.

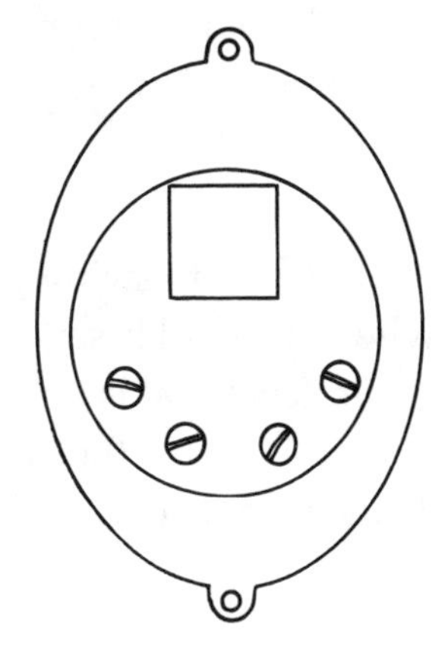

One common variation is for the plate to have the four standard lugs, plus a fifth lug in the center. If yours is like this, that center lug is virtually always a ground for safety.

Installing a new jack plate connector

Installing a new jack plate connector is a fairly easy job. The biggest problem is in routing the cable to that point (covered in the last chapter). Once the cable is brought to the site of the jack location, the final steps are straightforward in almost every case. We will begin with one of the most complicated methods—installing a wall jack that mounts to a stud beneath the wall.

It's important to note that there are a variety of wall box types available. Be sure that you specify that the box will be used in an existing wall. Some of these simply screw to the wallboard. Others, such as the one illustrated below, are more complicated to install but tend to be sturdier. If the one you get isn't like the one below, ask the salesperson for instructions on how to install the box.

Step 6-1. Locating the wall stud.
There are several different kinds of stud finders. All of them work on the same principle. An electromagnetic field is produced, which moves easily through the wall covering. The field reacts with the nails beneath, which is where the studs will be. (This method will *not* work if the wall is made up of a wire mesh coated with plaster. The device will "read" all the wire mesh. If you have this kind of wall, your best option is to leave the work to a professional or to use a surface-mounted jack).

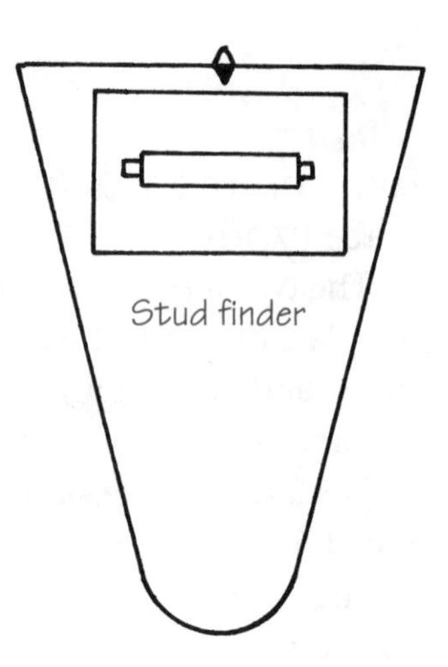

It is best to swing the detector slowly back and forth across the portion of the wall where the stud is several times and in several spots. You can then determine very accurately what is behind the wall.

Studs inside the wall are usually 1½ inches in width with the centers being 16 or 24 inches apart

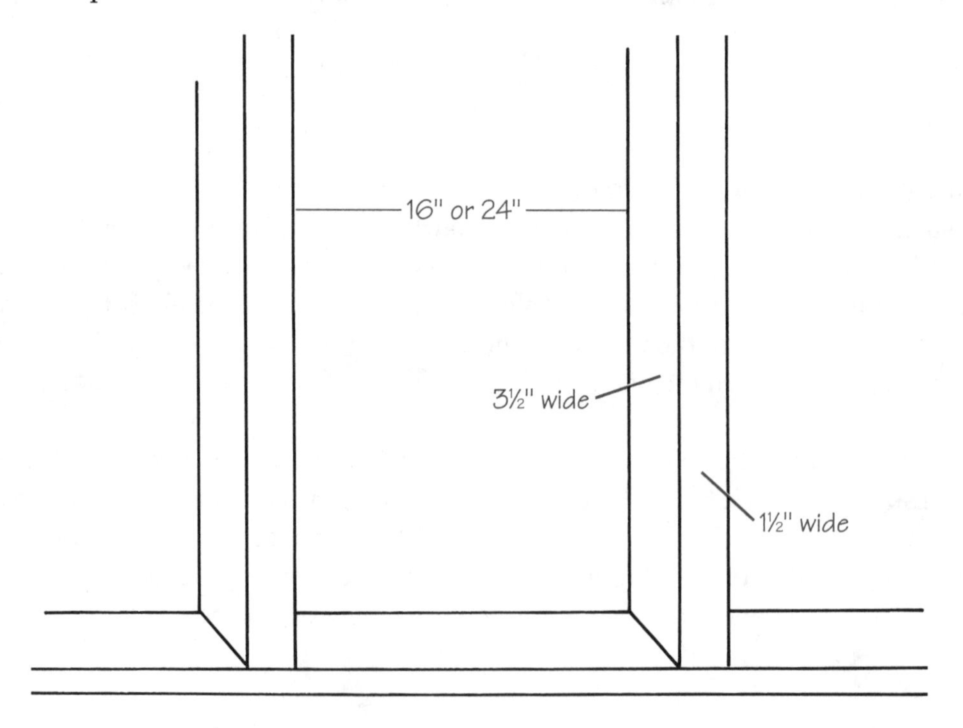

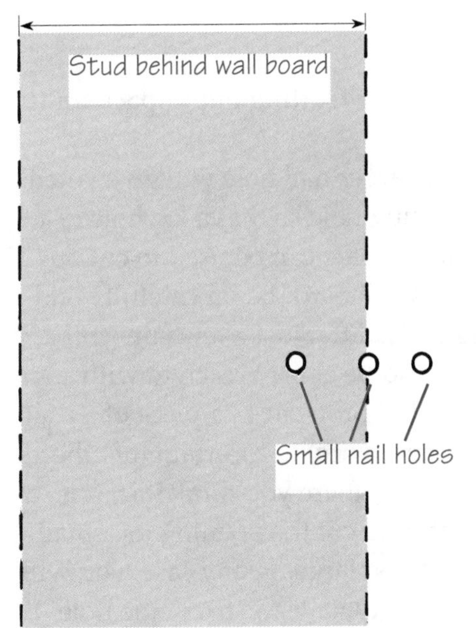

Step 6-2. Finding the edge of the stud.
Lightly pound a nail hole or two in the wall to determine the edge of the stud. You can tell if you're pounding into the stud because it will be a harder surface to pound into. Once you find the point that offers no resistance, you've found the edge of the stud(hopefully). Make this hole with no stud behind it a little large to give you a starting place to cut.

Step 6-3. Tracing the receptacle box.
Lay the receptacle box next to the stud and trace its outline with a pencil on the wall.

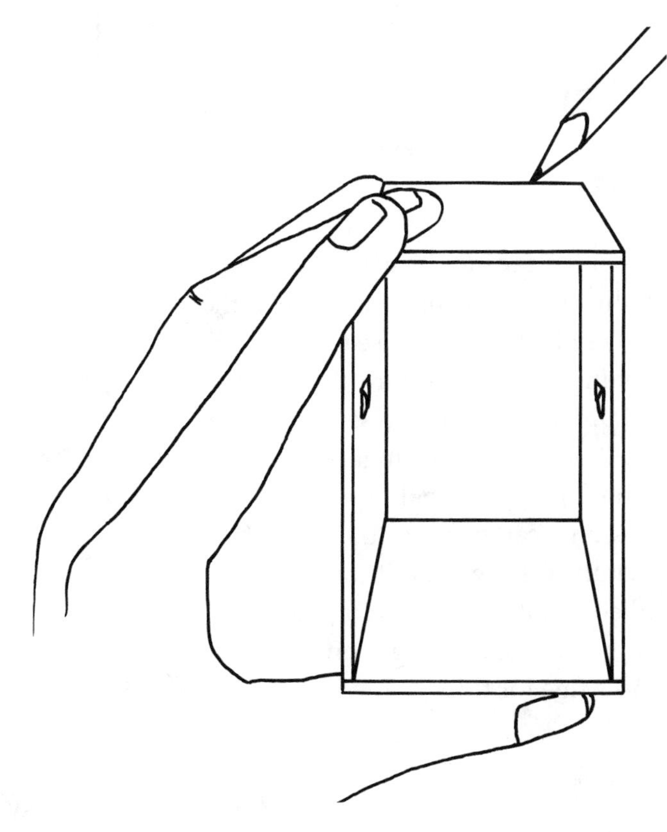

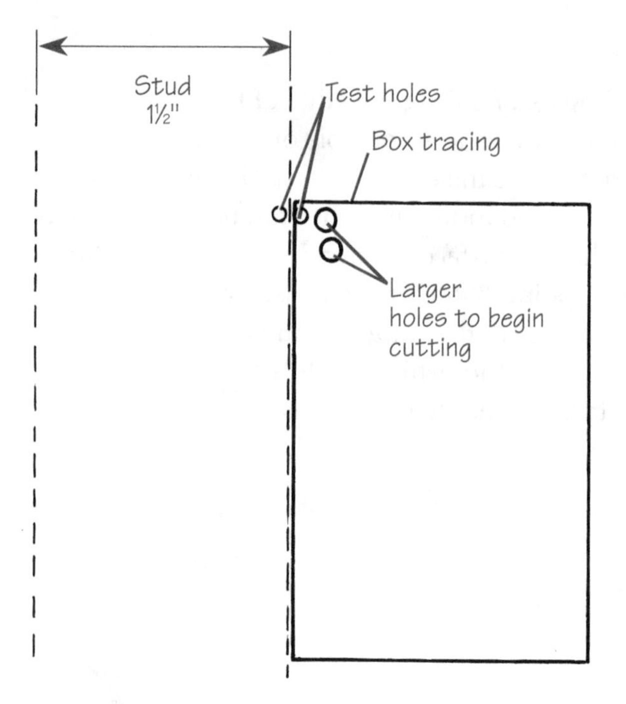

Step 6-4. Cutting out a space for the box.
The larger nail hole will give you a starting hole so that a keyhole or similar saw can be used to cut out the wallboard. Begin carefully and slowly. After just a few cuts you should be able to reach in with some kind of probe (e.g., a piece of a coat hanger wire) to ascertain that the stud is where you think it is. It is better to cut the opening too small than too large. In any case, you will probably have to "trim" the hole, such as with a sharp knife, to make a good fit that isn't too large.

Step 6-5. Feeding the cable through the box.
By this time the cable should be in the wall void and fairly close to the opening. If it comes from above, you will probably be able to see it and can pull it out with your fingers. If the cable is coming up from below, a wire with a hook on it can be used to "fish for"and catch the cable so that it can be pulled through the opening. A straightened coat hanger often works fine for this.

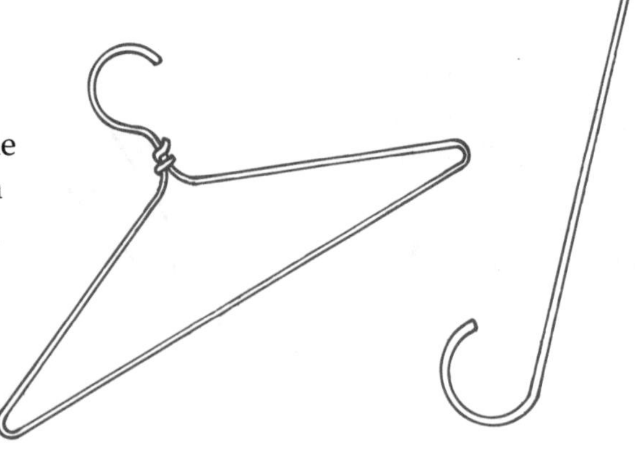

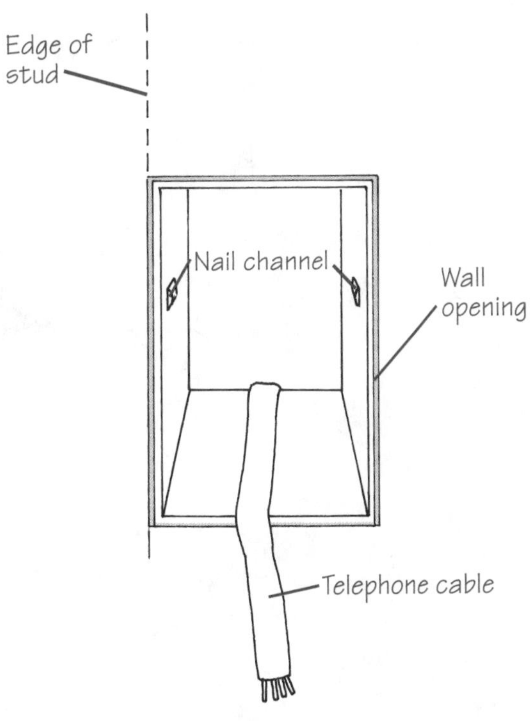

Step 6-6. Inserting the box into the hole. After pulling the cable through the box, push the box into place through the opening. (If you have followed the advice above, trimming the hole larger will be necessary.)

Step 6-7. Securing the box. The box is held in place against the stud with a nail that goes in at an angle. One or more holes/channels will be provided for this. This side is placed against the stud, and the nails are driven into the stud with a hammer.

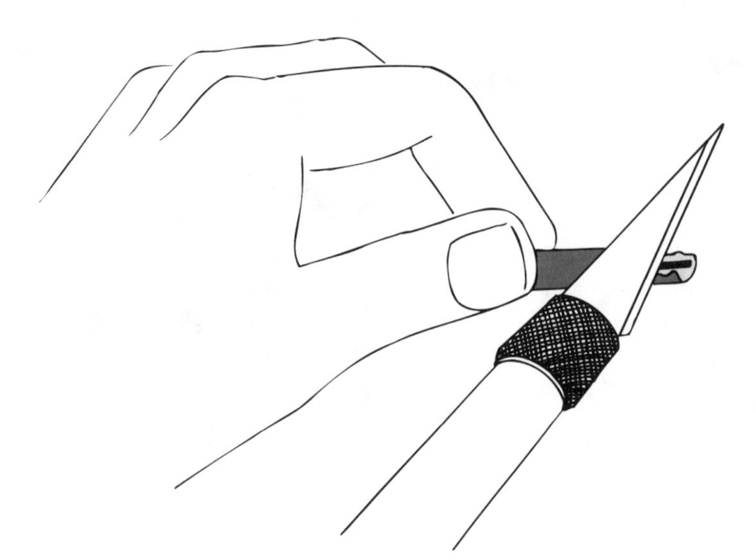

Step 6-8. Stripping the cable.
Using a sharp knife, strip off the outer cable insulation. Separate the wires inside, then strip off the insulation of those individual wires you intend to use.

Step 6-9. Connecting the wires.
Paying close attention to the correct color coding, place the bared wires beneath the screw lugs. It is best to form the wire into a hook first, with this hook going in the same direction as the tightening of the screw (clockwise). This helps to make a secure connection.

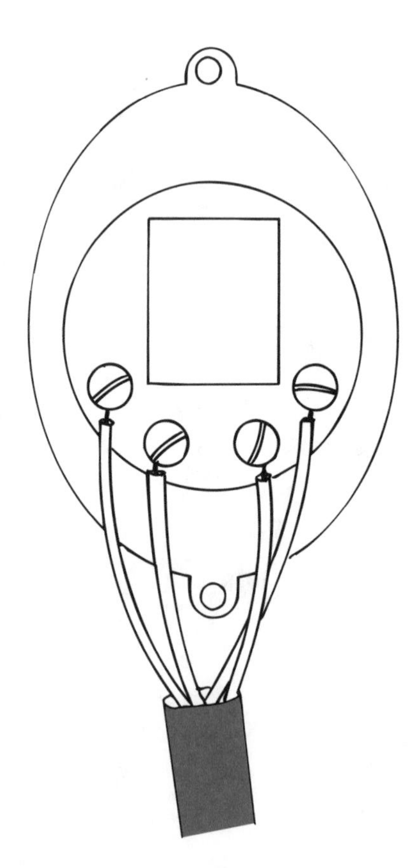

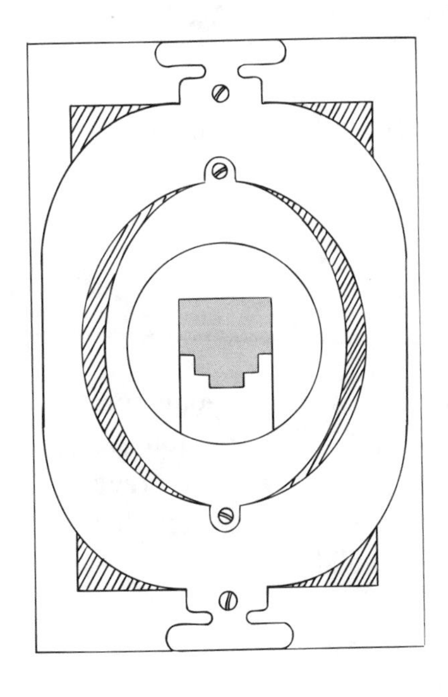

Step 6-10. Installing the jack plate.
The jack plate attaches directly to the box with two screws.

Sometimes the wall jack will have a separate cover plate. This is particularly true when the jack is meant to hold a wall-mounted phone. In this case a sturdier plate will attach to the box, and the cover plate finishes the installation. It attaches to the jack plate with two screws. (If the jack is meant to hold a wall-mounted phone, there will be studs that fit into the mounting plate on the back of the phone. If this is the case, the phone will have a movable plug on the back of it, which allows you to plug it in with the stud heads going into the circular slots. Then pull down to secure the phone into place.)

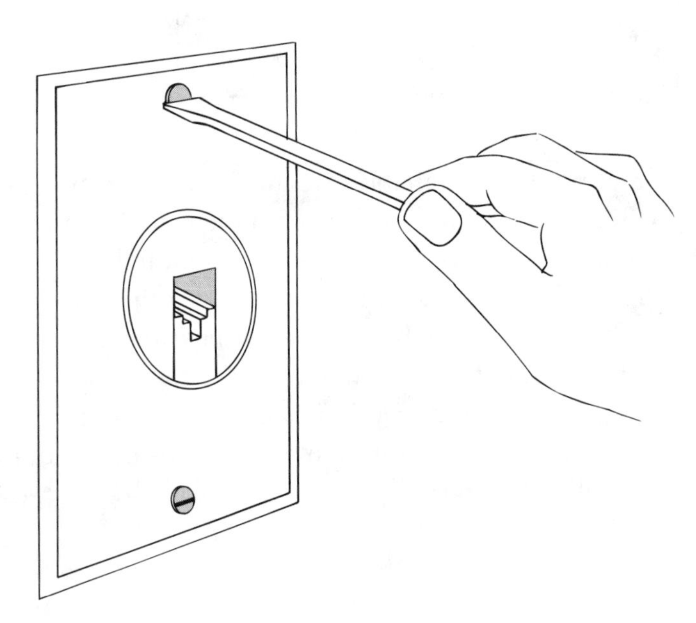

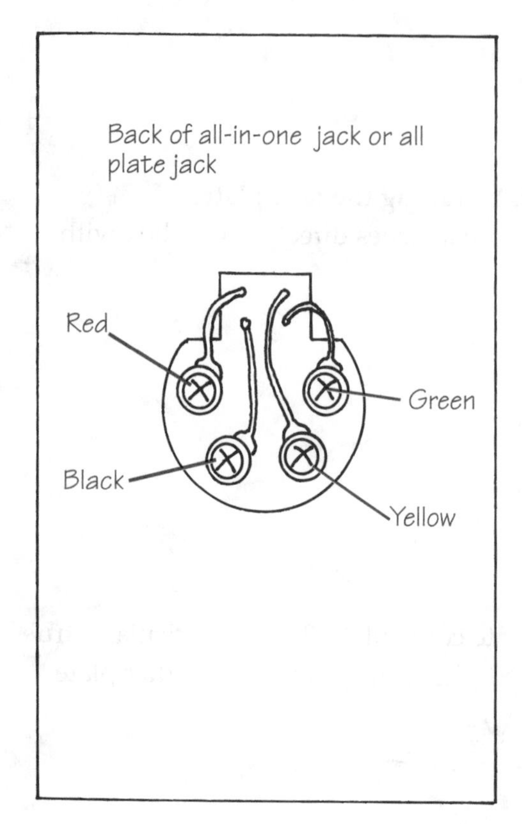

Often the jack is an all-in-one unit. These units do not have a separate jack plate. The wires attach beneath screws in the back of the cover plate, which then screws into the wall box.

Other types of installation are much the same but easier, because you won't have to find a stud beneath the wall. You might not have to cut into the wallboard at all. A surface-mounted jack, for example, ignores all stud-finding and wall-cutting steps (other than the small hole that might, or might not, be needed for the cable) are not necessary. Begin with stripping the insulation from the cable and from the wires within. Attaching this kind of jack usually involves nothing more complicated than putting a screw through a hole in the center of the jack and into the wall or baseboard.

CHAPTER SEVEN

The Telephone

There is actually very little you can do to troubleshoot a telephone, and less when it comes to repairs. This becomes even worse when the phone has the variety of services and features now considered common. Autodialing of at least the last number is almost standard. A more extensive memory with dozens of numbers stored isn't unusual. Very few phones, except for the "novelty" types, have a dial. They use a keypad that electronically generates tone pairs.

In most cases, the only truly mechanical part of the phone is the hook, which is the latch which moves up and down as you lift the receiver from its cradle. Most of the rest is electronic circuitry, with much of this consisting of components that can't really be tested—or replaced.

Still, there are some things you can do if you have a phone that is acting up. These steps will also help to verify that the problem is in the phone—or if it's in the line.

Step 7-1. Trying a different phone

As has been pointed out elsewhere, the first and easiest step is to try a different phone in that same outlet. It's possible that two phones have gone bad at the same time but it is unlikely. If the second phone works, you instantly know that the first phone is bad. If you don't have a second phone you know to be good, try using the same phone in a different jack. This won't tell you much, however, unless you already know that this second jack is good. Consider borrowing a neighbor's phone that works and plug it into your outlet, or plug your own phone into a working jack in your neighbor's home.

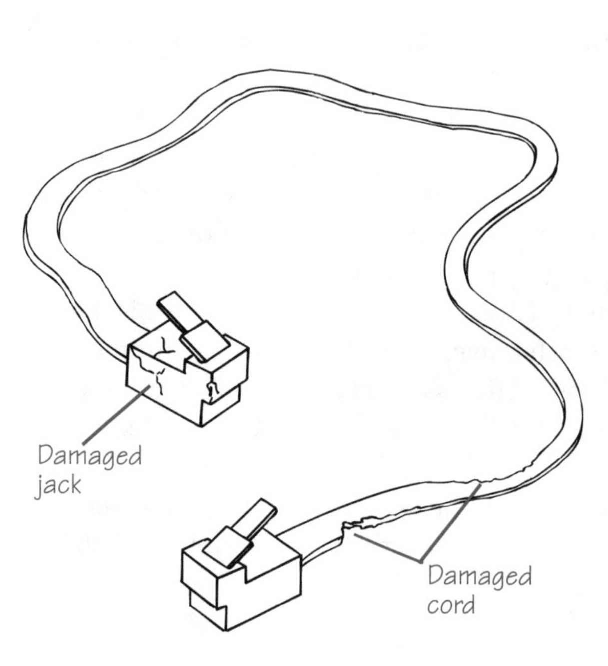

Step 7-2. Examining the connector and cord.

Sometimes the only problem is that the cord between the phone and the jack has been damaged. Often a visual examination will reveal the problem. Look closely at the fine wires of a modular plug (and those inside the jack if two phones fail to work in that jack).

Step 7-3. Testing the cord with a VOM.

The easiest way to test a cord is by replacing the cord with a cord you know is good. Unfortunately, not many of us have spare cords lying around (although you could substitute with a cord from a phone known to be working). Lacking an extra cord, it's fairly simple to test a cord with a VOM. The details on how to do this are found in appendix A.

A continuity check, to be sure that the wire is continuous and unbroken, calls for probing that same wire on each side of the cord. Unplug the cord from both sides and carefully test each wire in turn. The reading you get should be zero ohms.

See step 4-6.

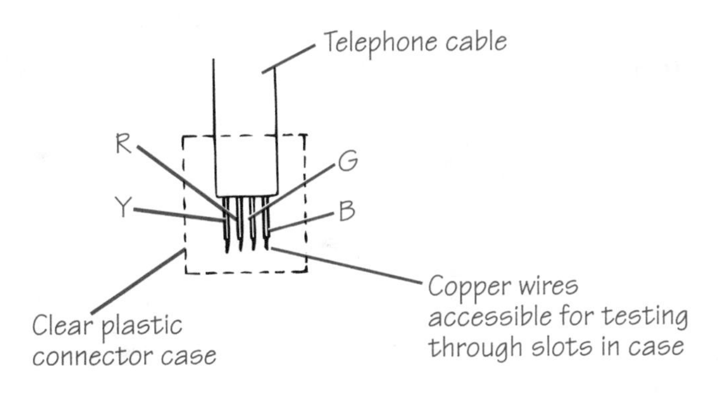

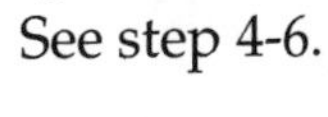

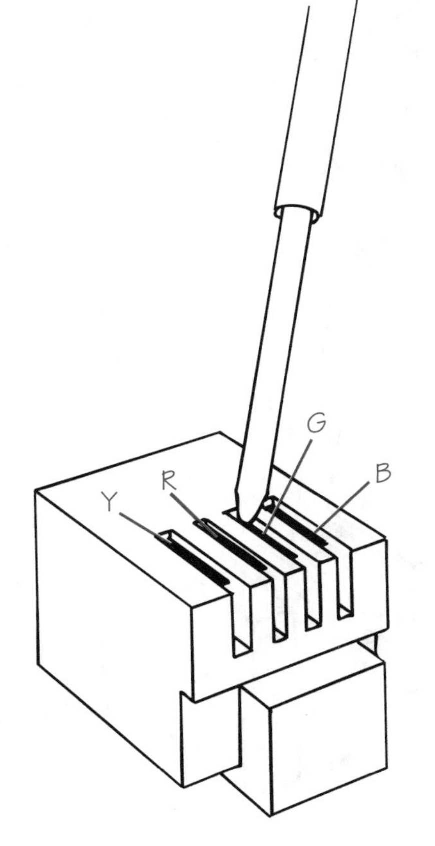

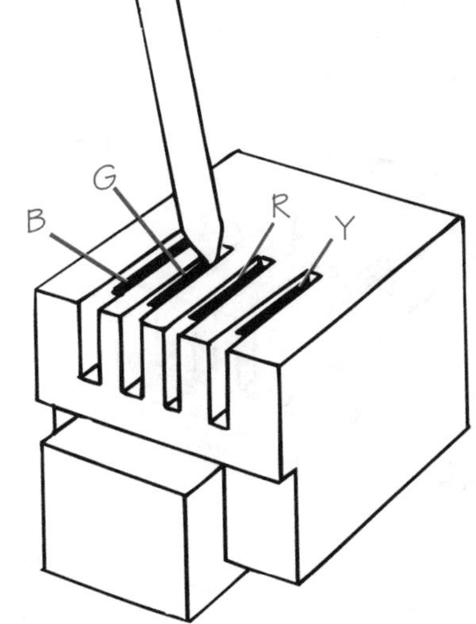

Y=Yellow
R=Red
G=Green
B=Black

Testing for shorts, wires touching each other inside the cord, takes a little more time. Touch one of the probes to a wire and the other probe to each of the other wires in succession. The reading should be infinity (no scale movement), which shows that there is no contact between the wires. Again, refer to step 4-6 for further detail.

These same tests are used to test the cord between the base and the handset. In both cases, if either test fails the cord is bad and must be replaced.

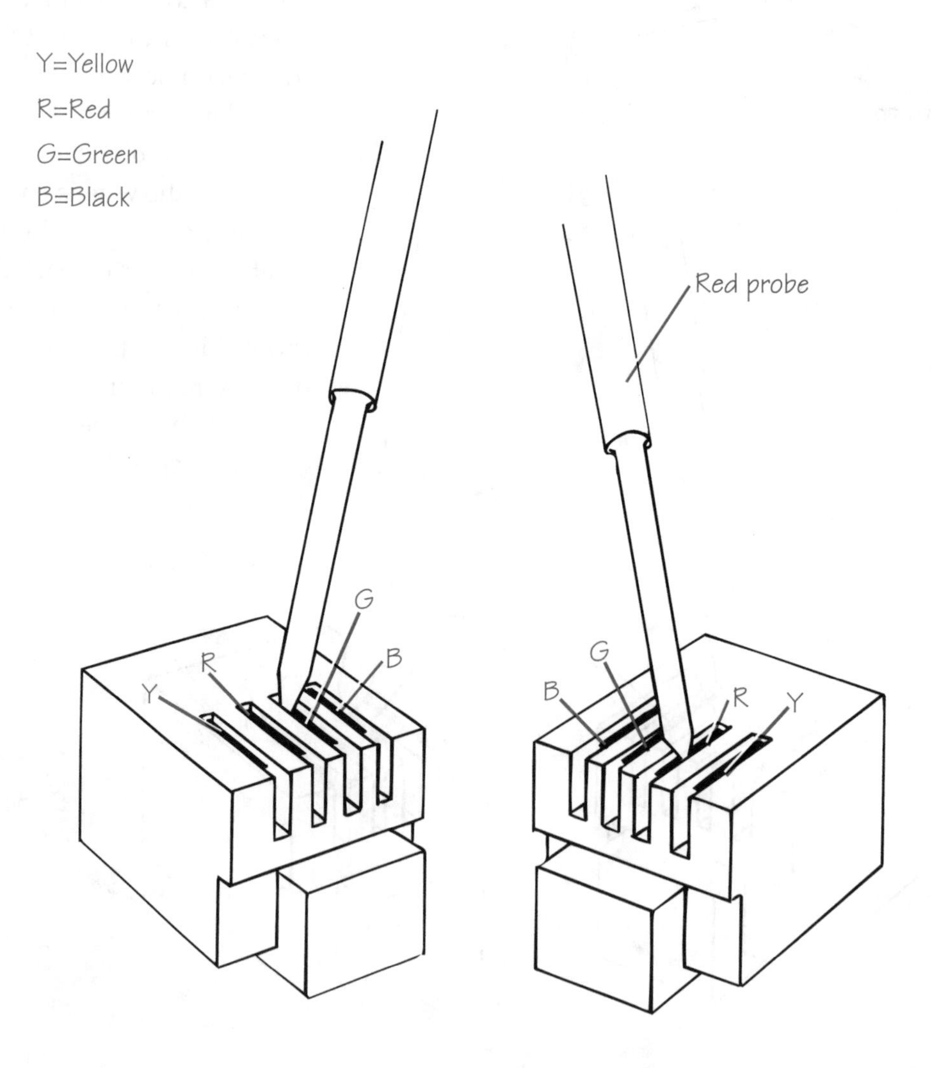

Step 7-4. Opening the phone.

If you have determined that the jack is functioning and that the cord is good, you have also determined that the problem is in the phone. To proceed with troubleshooting you must open the phone. Exactly how you do this depends on the kind of phone you have. The steps you follow after you open the phone remain the same. Mostly it involves a visual examination, looking for obvious signs of damage or corrosion.

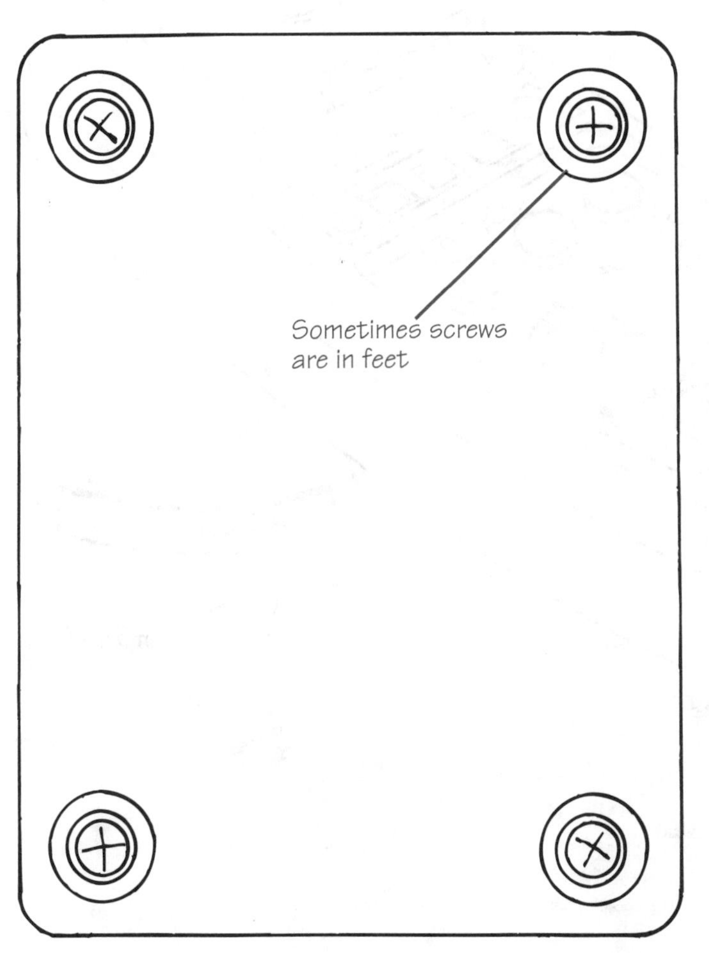

The standard desktop phone has a plastic top cover that fits over the rest of the phone. It will be held in place by screws (almost always four: one in each corner) through the bottom. With this cover removed, you can get at almost everything inside the unit.

The handset holds much of the electronic circuitry, including the tone pad. Without some fairly sophisticated equipment you can't test much here, but you can verify that there is no physical damage.

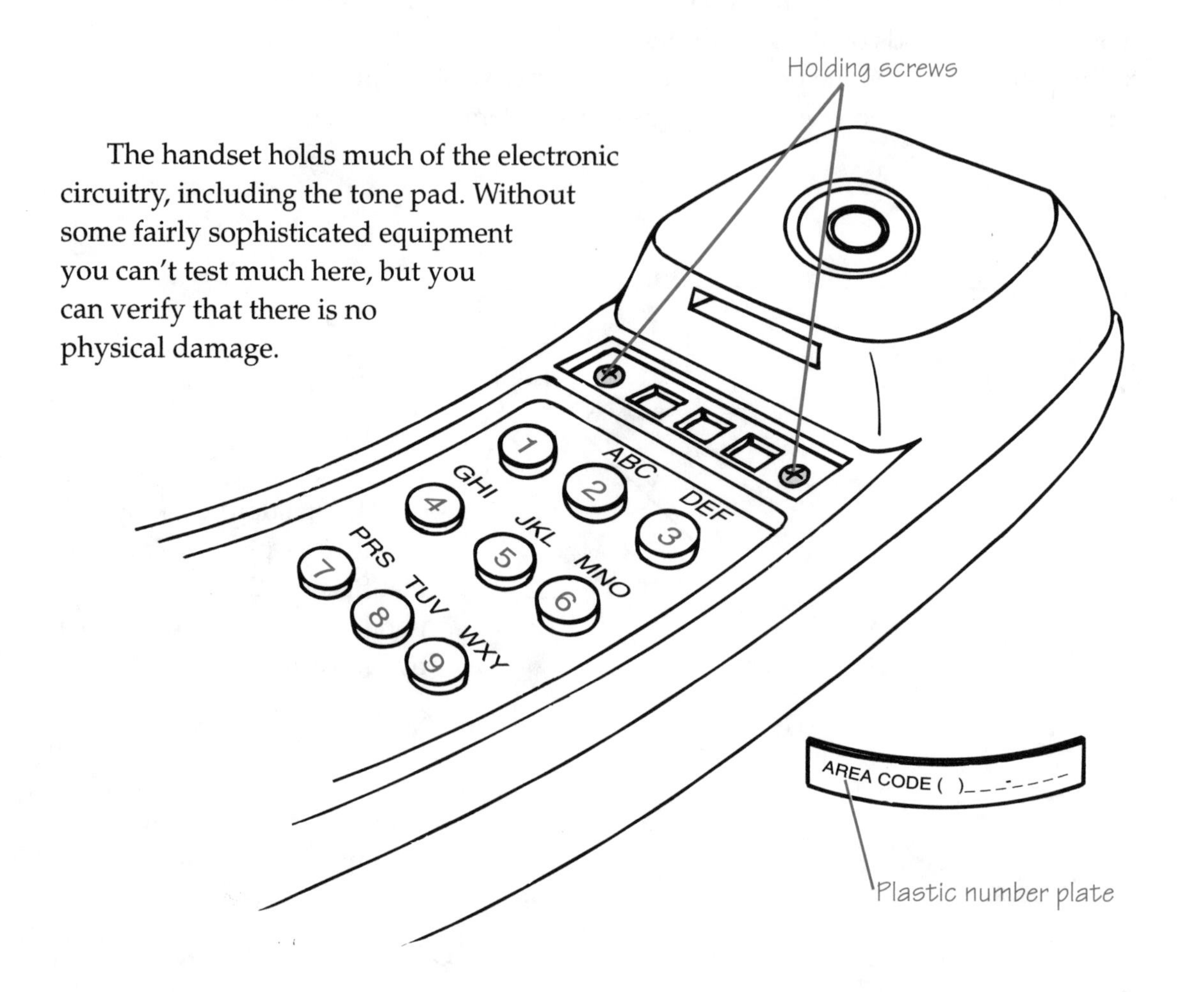

The base is often held together by four screws: one in each corner beneath the base (the same as with other desk-type phone).

Sometimes, especially with wall-mount units, the base is held together with a single holding screw hidden beneath the number plate. The small plastic cover and paper tag come out easily by prying on one side with a small screwdriver. The screw beneath the plastic cover can then be removed, and the base should come apart easily.

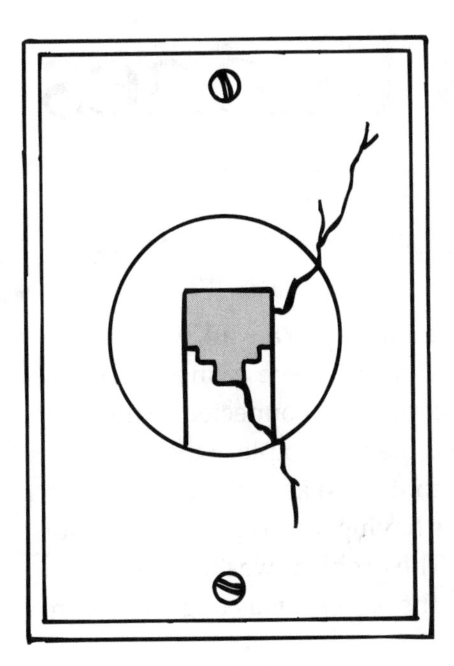

Step 7-5. Examining the phone. Visually examine the phone for signs of damage. Pay particular attention to wires that might have broken or come loose. Especially if the phone has been yanked, it's possible that the jack has broken one or more of the connections.

CHAPTER EIGHT

Cordless Phones

The first telephones were mounted firmly to a wall, with the mouthpiece on the wall unit and the earpiece connected by a short cord. Next came the "desk phone," connected to the incoming line with a wire that enabled the user to place the phone more conveniently. Variations of this continued as the telephone system developed, with the phone sometimes having a cord long enough for the person to walk out into the yard. The problem was that the phone still used a cord. The longer the cord, the easier it was to get that cord tangled, tripped on, cut, etc.

The solution was the cordless phone. It became so popular so fast that neighbors were soon inadvertently tapping into each other's phone lines. That brought on a selectable security switch and eventually a complete redesign of the local system.

Whichever type of unit you have, the basic principles are the same. A base unit connects to the telephone company's lines and supplies power to the local cordless system. The handset is the user's means of communication. Both base and handset are transceivers, meaning that they transmit and receive signals.

Problems with the telephone lines in homes are often caused by cordless phones. This is why when you call the telephone company because of a problem, one of the first things a repair man will ask is if you have a cordless phone. For this reason, you should disconnect the cordless phone from the telephone jack first when you begin to troubleshoot.

The symptoms vary. There might be crackling on the line, the line might be completely dead, or almost anything between. If the symptom clears when you unplug the cordless phone, you've found the cause. Now all you have to do is fix it.

Cordless base unit

The base unit is responsible for connecting the cordless handset to the rest of the world. It's also responsible for providing the electrical power needed for keeping the handset and base in communication.

To do this, the base plugs into the wall outlet. This goes into the power supply where it does two basic things. The first thing it does is power the transmitter and receiver in the base. Its other function is to recharge the batteries in the handset.

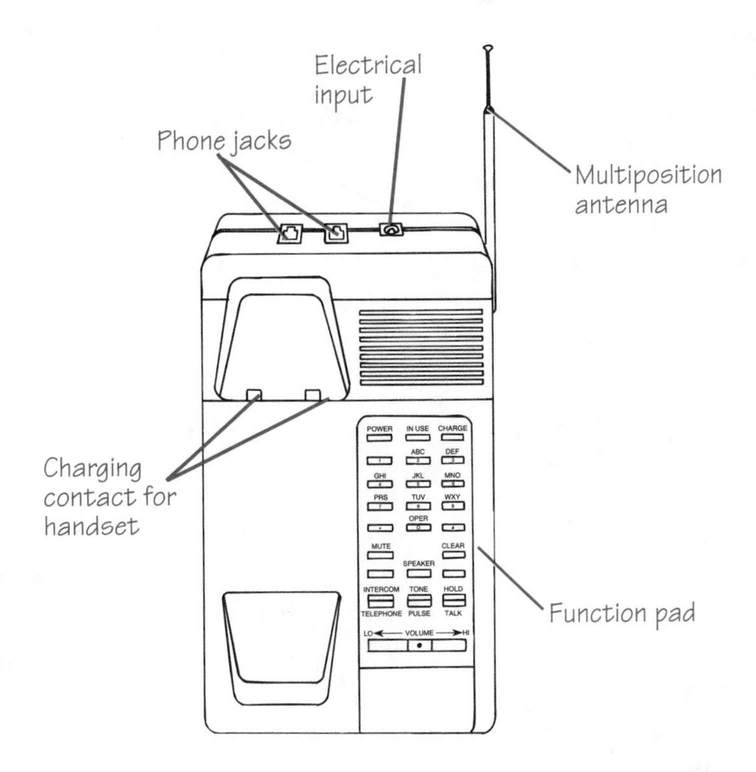

The base also connects to the telephone jack. Testing this part of the system is the same as with any telephone. That is, if the cordless phone isn't working, check the jack and the cable that connects the base to the jack.

Many times the base will also have a "page" feature that sends a beep to the handset. This can be used to signal the person who has the handset, and it also serves to locate the handset when you misplace it.

Cordless handset

The handset is powered by rechargeable batteries. These batteries can be held in a compartment, making them accessible to the user, or they can be installed inside a more sealed handset. Either way, a pair of contacts on the handset touch against the charging pins of the base. Each time the handset is placed back into the cradle of the base, the charging system becomes active and the batteries are "filled." These batteries provide the power needed by the receiver and transmitter in the handset.

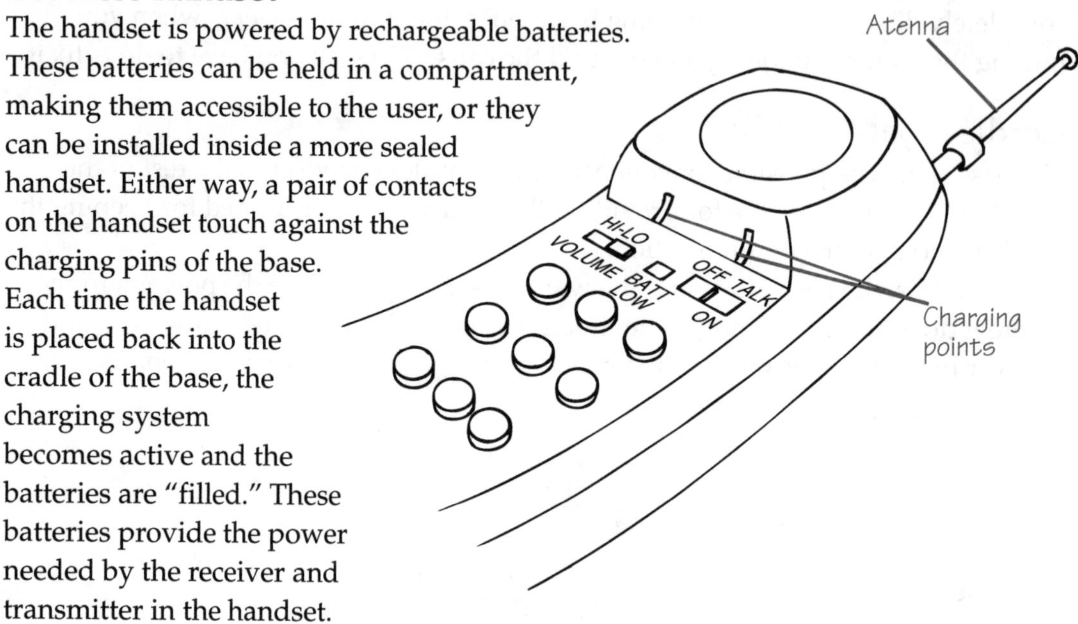

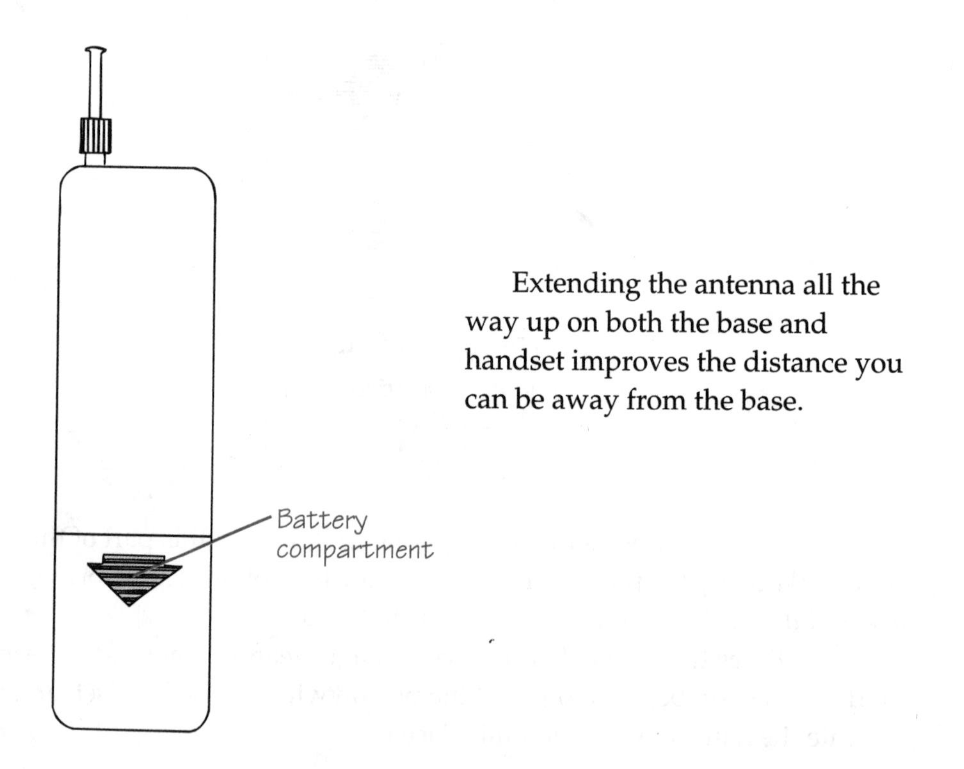

Extending the antenna all the way up on both the base and handset improves the distance you can be away from the base.

Battery packs

The battery pack is almost always made up of several individual cells that are wired together and then sealed into the package form. The most important consideration when replacing a battery pack is to be sure that the voltage and available current (amperage) matches. Both will usually be clearly stated on the label. Unless you want to strap the pack onto the handset with tape, it must also be of the proper configuration to fit inside.

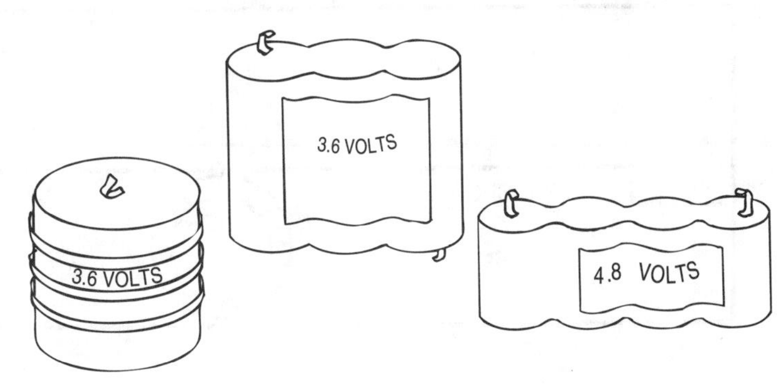

Step 8-1. Checking for incoming power.

The base unit plugs into a wall outlet. If nothing at all is happening, check to be sure that the base is plugged in, and that the outlet is good. This can be done by plugging something else into that outlet or by probing it with the VOM (see appendix A). Almost every base will have some kind of indicator light that shows that the handset is charging. (This same light sometimes indicates that the handset is in use.)

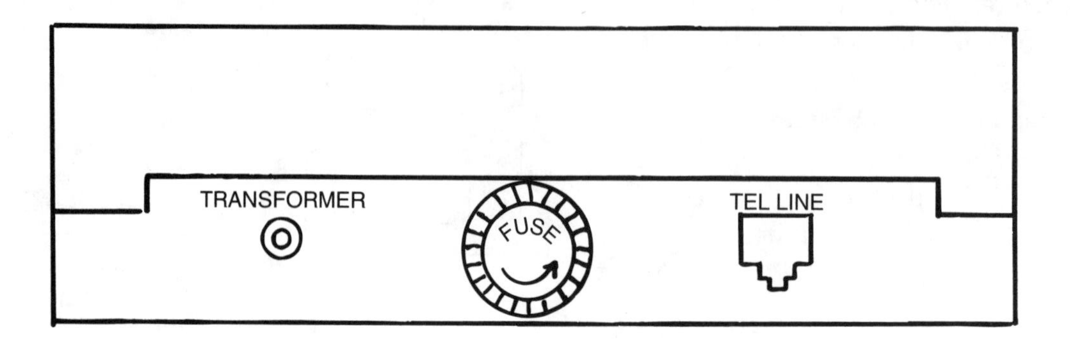

All base units will have a fuse to protect the unit. This fuse is often located inside the base and is not easily accessible. However, some base units have a user accessible fuse, usually beneath a screw cap in the back.

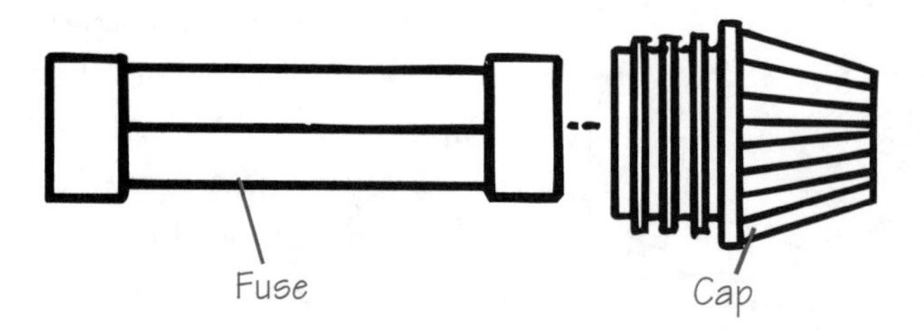

If there is no power to the base, unplug the unit, remove the fuse, and examine it. Usually you will be able to see if it is good or if it has blown.

Shopping List for All Thumbs Guide to Telephones and Answering Machines

Basic

❒ Phillips screwdrivers
(A medium-sized head—about 1/4 inch—and a small-sized head—about 1/8 inch.)

❒ Blade screwdrivers
(Same sizes as the Phillips screwdrivers.)

❒ Needle-nose pliers

❒ Regular pliers

❒ Nutdriver set

❒ Volt-ohmmeter

❒ Working telephone

❒ Caulking

❒ Stapler

❒ Clamps

❒ Drill

❒ Signal generation
(See Appendix B.)

❒ ____________________

❒ ____________________

Blade screwdriver

Phillips screwdriver

Needle-nose pliers

Regular pliers

Volt-ohmmeter

Clamp

Stapler

STAPLES
CHISEL POINT

7 PC NUT DRIVER SET

Nutdriver set

VARIABLE SPEED
REVERSIBLE
DRILL

Drill

COVERS WITH PAINT
CAULK

Caulking

Working telephone

Cleaning

❒ Isopropyl alcohol
(At least 95 percent pure.)

❒ Foam swabs/chamois

❒ Cleaning fluid

❒ ____________________

❒ ____________________

Isopropyl alcohol

Foam swab

Cleaning
solution
Non-residue

Safety Tips

Before you start

❒ Think safety.

❒ Read all the instructions carefully.

❒ Gather all your tools and required materials.

❒ Make sure your test equipment is working properly.

❒ Work slowly and carefully.

Limitations

The most important factor is safety—yours first and then the safety of the unit or system you are working on. Nothing is more important!

The purpose of this book is to help you understand that you have fewer limitations than you might have suspected. However, you still must be able to recognize your limitations.

What if cleaning the machine doesn't cure the problem? You can try a few more things, as detailed in the book. Many times you'll be able to fix the problem, or at least spot the cause. Approach the problem with confidence, and realize that in most cases you can do it.

However, only you can tell when you've reached your limit. Be honest enough with yourself, both so that you don't stop too soon, and so you don't go too far.

Eventually the point is reached when you face the choice of calling a professional, or buying a new one.

If in doubt, set your VOM to read resistance in the R×1 range.

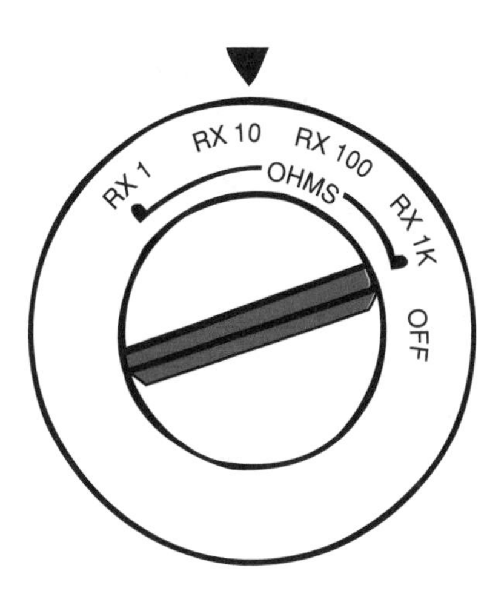

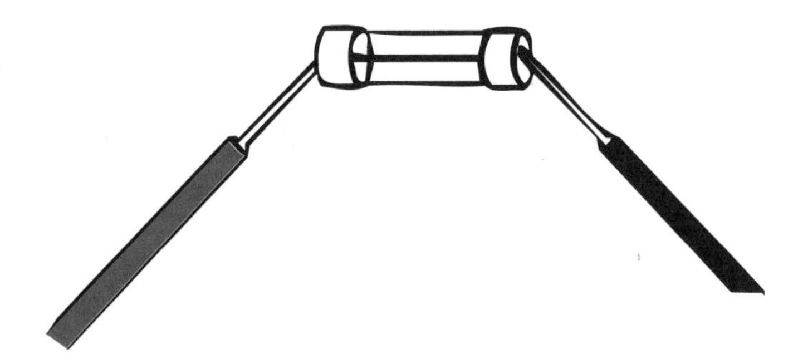

Test the fuse for continuity. (See Step 4-12.)

The reading should be 0 (zero) ohms. If the reading is high or infinite, the fuse is bad and must be replaced.

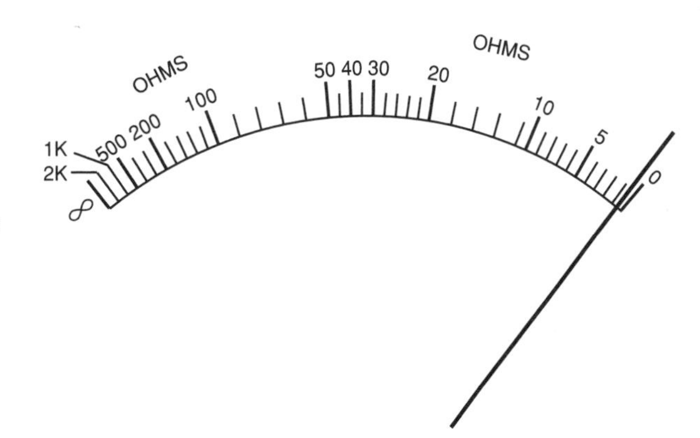

Step 8-2. Checking the charger.
Set the VOM to read dc volts in a range above and closest to the voltage listed on the battery pack.

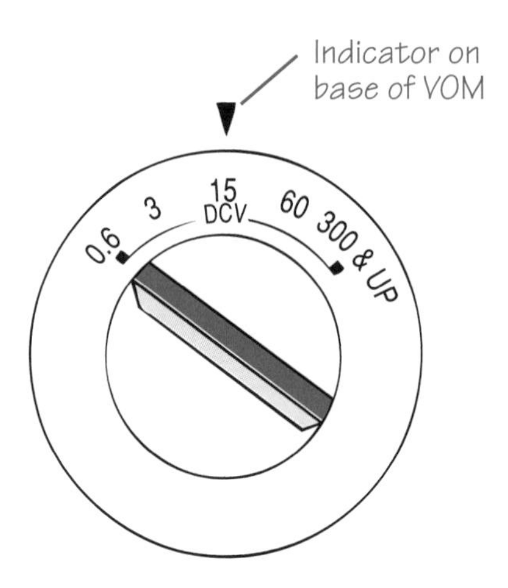

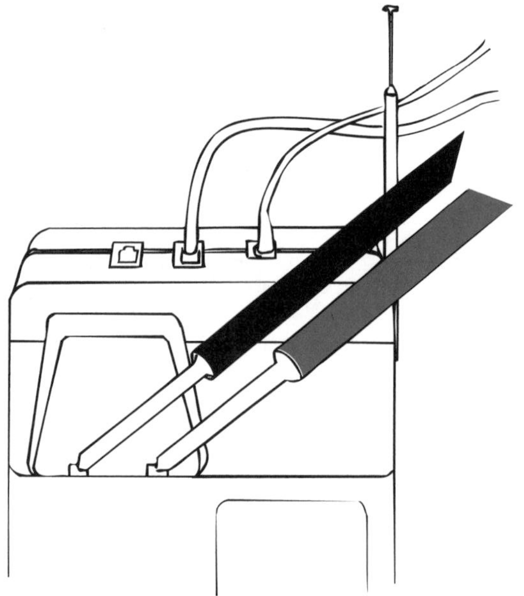

Touch the probes to the charging pins on the base. If the needle moves in a downward direction, reverse the way you are touching the probes to the pins.

The reading you get should be close to the voltage listed on the battery pack.

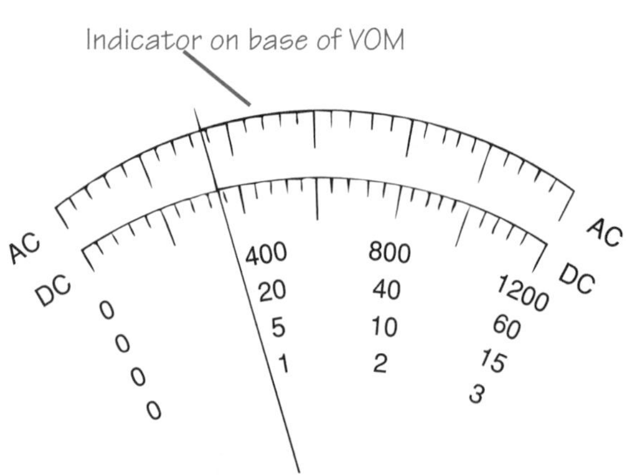

Step 8-3. Checking the battery pack.
Rechargeable battery packs usually have an expected lifespan of about 2–3 years, after which they must be replaced. One of the first signs of battery failure will be that it is not taking or holding a charge. Although this could be the fault of the charging system, it is much more likely to be that the battery is simply worn out.

Set the VOM to read dc volts in a range above and closest to the voltage listed on the battery pack.

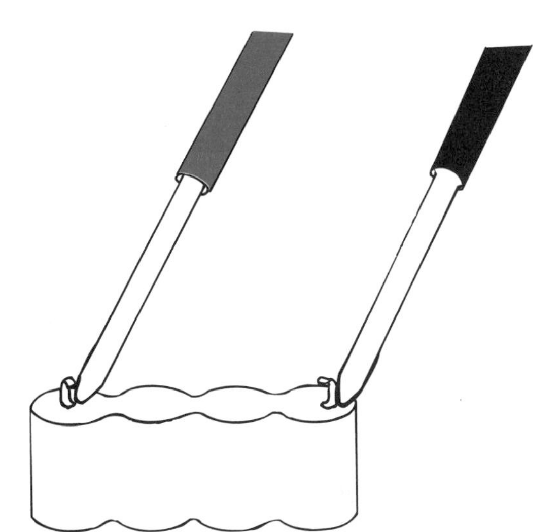

Touch the black probe to the negative terminal on the battery and the red probe to the positive terminal. The reading should be very close to the voltage listed on the battery pack. If it's not, the battery is not charged. This means that you either have a battery problem or a charger problem (which was checked in Step 8-2).

Remember that the replacement battery pack must be an exact match.

Step 8-4. Checking internal battery packs.
Some cordless handsets, particularly the older models, are sealed with the battery inside. This means that if the battery pack goes out, you have to open the handset to replace it. Often this process requires soldering. (It's common for an internal battery pack to be soldered into place, which means you have to desolder the old pack and solder in the replacement.

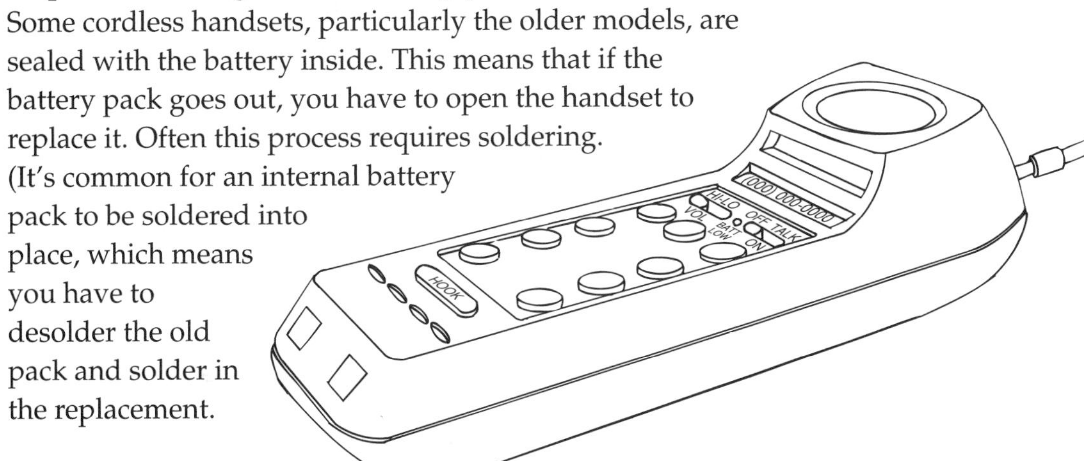

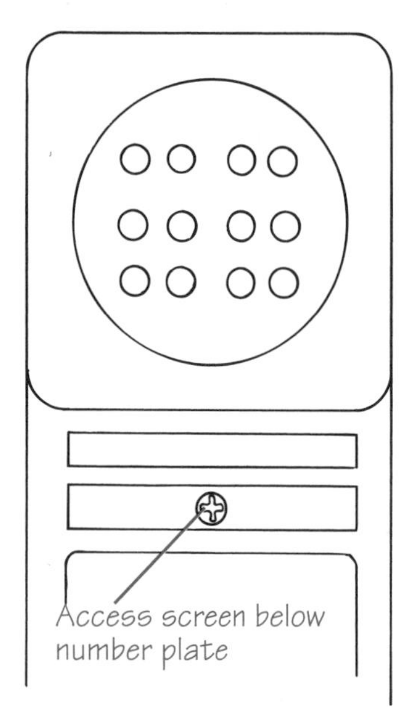

The handset is usually held together by screws. Usually the handset will have a single screw located beneath the number plate.

There will also usually be catches, often at the top and the sides toward the bottom. You might have to pry the case apart. If so, do this very carefully so that you don't damage the plastic of the case.

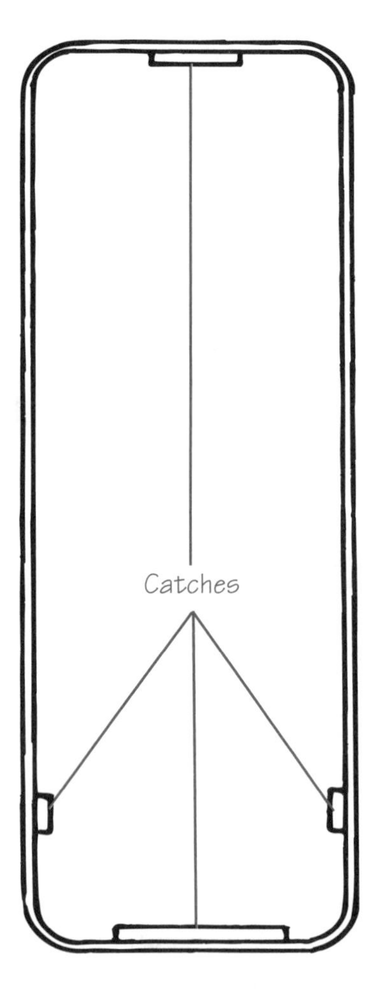

Bottom of sealed telephone showing catches

Many times there will be components in the back of the case, with wires running from these components to the main circuit boards.

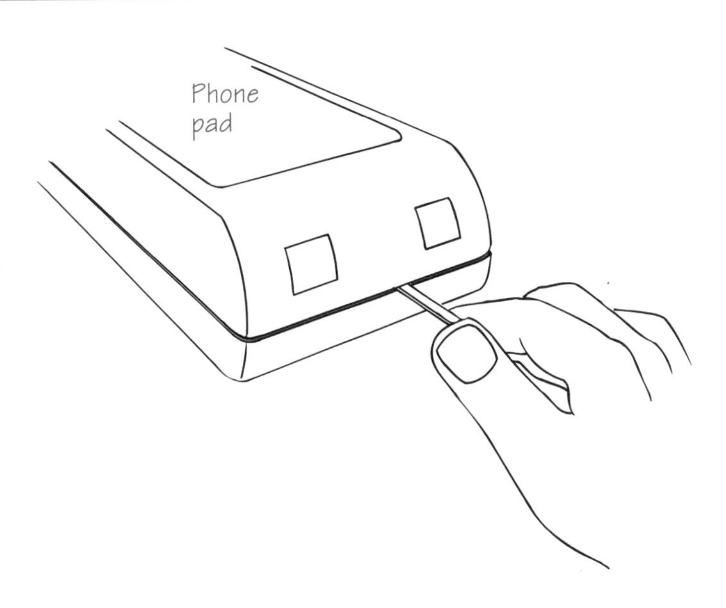

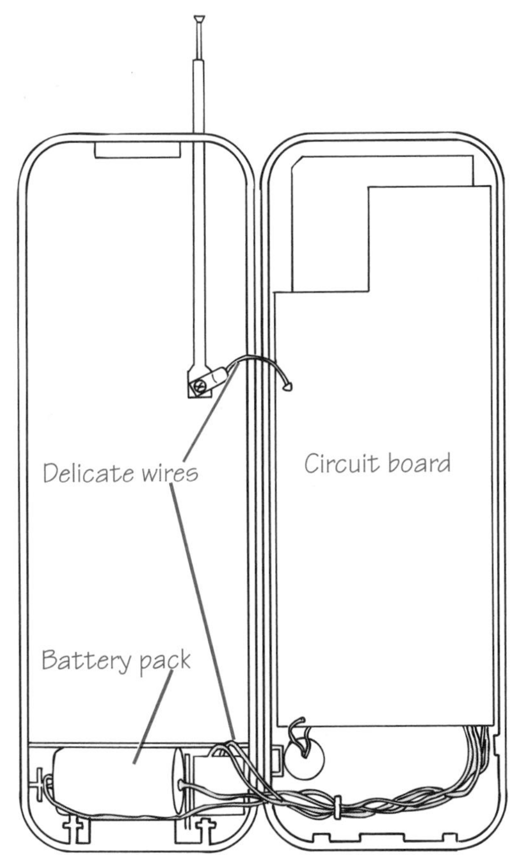

When opening the phone, be very careful to not break or damage these wires.

Locate the internal battery. Set the VOM to read dc volts in the range close to but above the voltage listed on the battery. (You might have to remove some padding to see this label.)

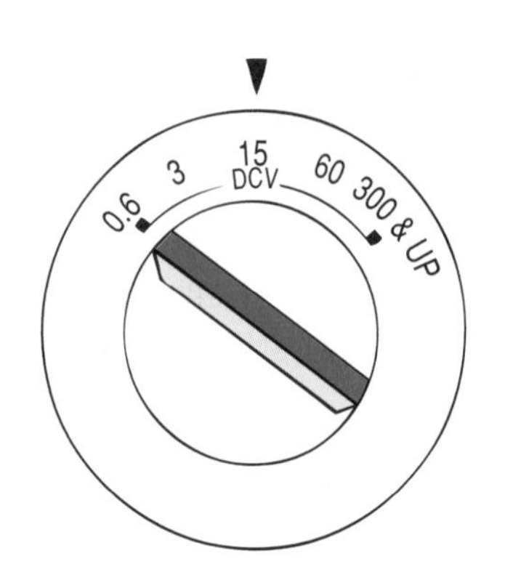

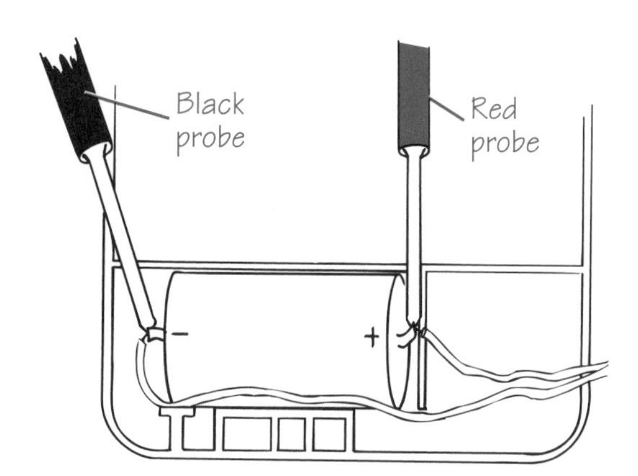

Probe the battery and check the reading. The meter should show the voltage to be very near the value on the label.

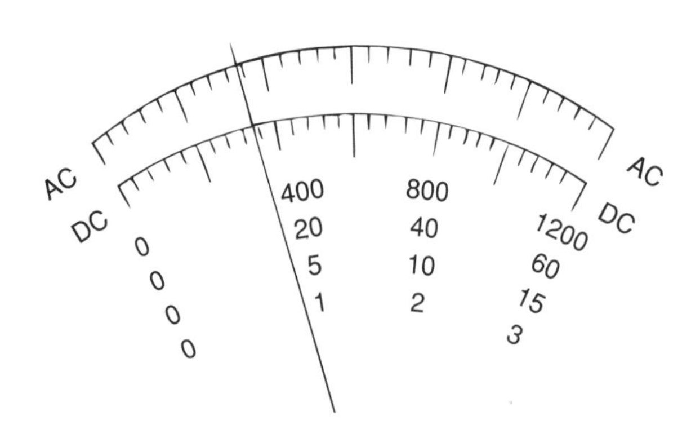

Step 8-5. Checking the "privacy code."
Most cordless phones today use either a channel or a privacy code so that the handset works with your base only. If the setting isn't the same on both the base and handset, the system can't work. Refer to your owner's manual to determine how this is set with your phone.

CHAPTER NINE

Answering Machines

Answering machines are common devices these days. There are many different models and types from which to choose, with the price range going from below $50 to several hundred dollars.

The machine uses audio cassettes, electronic memory, or both. The unit can be built into a telephone, or it can be a separate unit. The features available vary from model to model. Many offer a time and date "stamp" that automatically records the time when a call comes in. Also common is a hang-up detect so that no recording is made if the caller hangs up without leaving a message. (Older machines recorded for a certain period of time whether a message was left or not.) Another kind of sensor (usually called VOX, which stands for "voice activated") records only as long as the caller talks and automatically shuts off the recording when the call is over. There might be a switch to set the maximum recording time allowed, such as 30 seconds, 1 minute, or 5 minutes. Virtually every answering machine has a switch to set when the machine will answer (first ring, fourth ring).

The more features the unit has, the more electronic it is. To the average end user, this means that there is less that can be done to troubleshoot and repair the unit. For example, units that are "all electronic" have no mechanical parts—no cassettes. When something goes wrong with such a machine, you are very limited in what you can do other than to check to see that the unit is getting power.

Most answering machines have at least one cassette. If there is only one cassette, it is used to record incoming messages. A machine with two cassettes has one cassette for the outgoing message and one cassette for those coming in. In either case, the cassettes are one of two standard sizes: regular (like an audio cassette) and micro (a smaller version that enables the answering machine to be smaller).

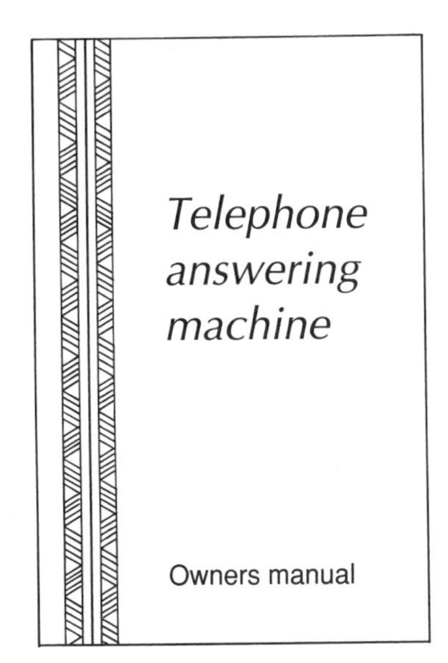

Step 9-1. Reading the manual.
Operating a telephone, even a cordless telephone, is simple. The same is not always true when it comes to using an answering machine. The more features your answering machine has, the more complicated the operation is. You should be thoroughly familiar with the manual for your machine. In fact, the most common "malfunction" with answering machines is simply that the owner isn't using it correctly.

Learn the machine's features and how to use them. This is often the only way you'll know if the answering machine is truly malfunctioning or not.

Step 9-2. Checking for power.
Most home answering machines, and many other telephone accessories, get power from a transformer which converts the nominal 117 Vac to the needed value (usually 22–24 Vac). Sometimes the transformer needed is specific to the machine. If your machine is completely dead, first make sure the transformer is plugged into a good outlet and that it is the right transformer for that machine.

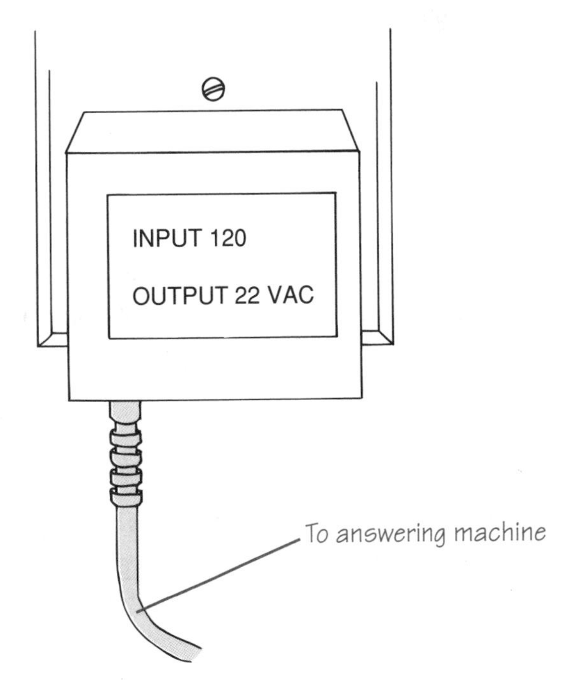

Unplug the power connector from the answering machine. Set the VOM to read volts ac in the proper range (as per the label on the transformer).

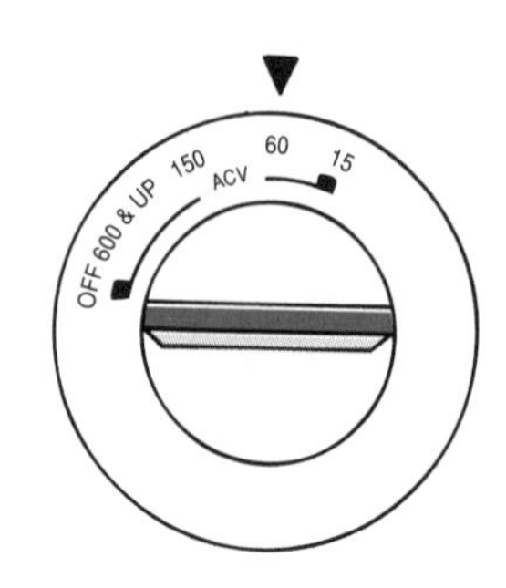

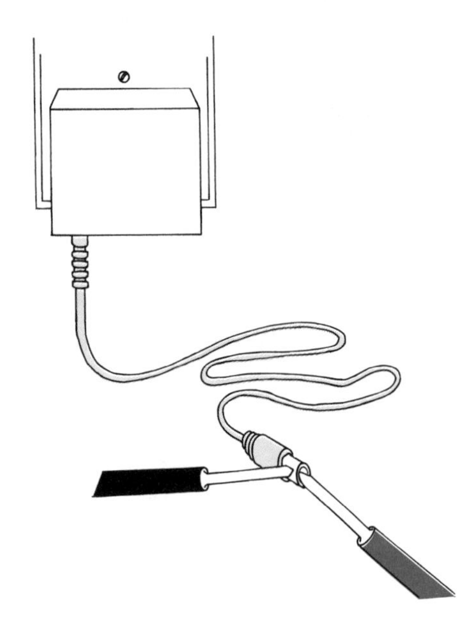

Touch the black probe to the outer shell and the red probe to the inside of the connector.

The reading should be very close to the output rating of the transformer. If it's not, the transformer is bad and must be replaced.

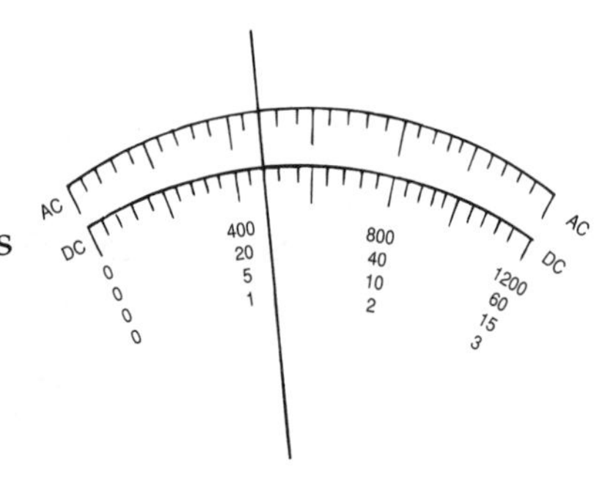

If the answering machine is working fine but it keeps losing electronically stored information (time, date, etc.), it's possible that your area has suffered from power outages and that the back-up battery (if your unit has one) is dead. In most cases this is a single 9-volt battery. Another fairly common method is to use two or more AA or AAA batteries.

Step 9-3. Checking the battery.
The compartment for the battery (or batteries) is usually under the machine. Find this and slide the door open. (If you can't find the back-up battery compartment, refer to the answering machine's manual to find out where it is, and if the unit even has this feature.)

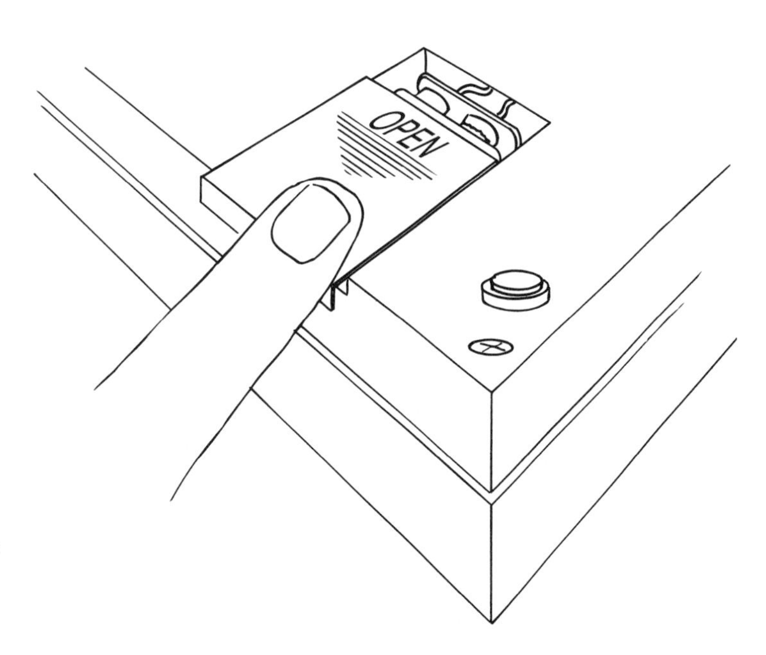

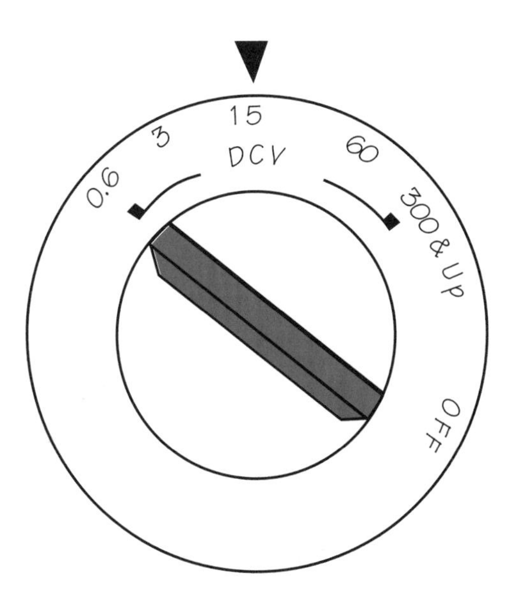

You can remove the battery for testing. To get the most accurate reading, leave the battery in place and shut off the machine. This places the battery "under load."

Set the VOM to the appropriate dc voltage range. Touch the red probe to the positive terminal and the black probe to the negative terminal. The reading you get should be very close to the full battery rated charge. (If your machine uses AA or AAA batteries, each battery holds 1.5 volts. Therefore, two such batteries would provide 3 volts; three batteries would yield 4.5 volts, and so on.)

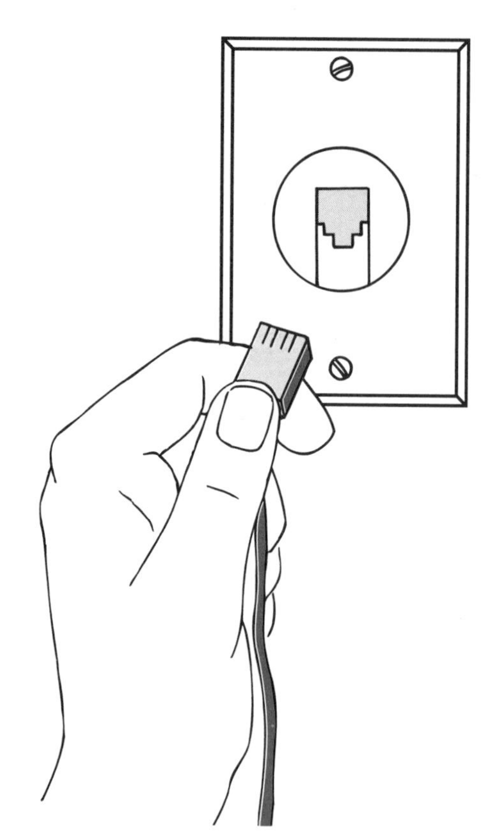

Step 9-4. Checking the phone connection. As with any phone device, the answering machine must be connected to a good phone jack through a good phone cord. The tests for this are the same as they were for the telephone itself. Is the unit plugged in? Is the outlet good? Is the cable good? Has the answering machine been damaged in some way? (This last one requires that you open the machine and examine it for damage.)

After you've examined these possibilities, you've done pretty much everything you can do if your answering machine is completely electronic. Troubleshooting beyond this point requires that the answering machine be tested in more depth, and with fairly extensive training and equipment.

If you have a unit that uses cassettes, however, you can continue to examine your machine.

Step 9-5. Examining the cassette. If recording or playback (in a dual cassette machine) is erratic, examine the cassette for damage. (Don't forget that the cassette might be failing to record because it is already full.)

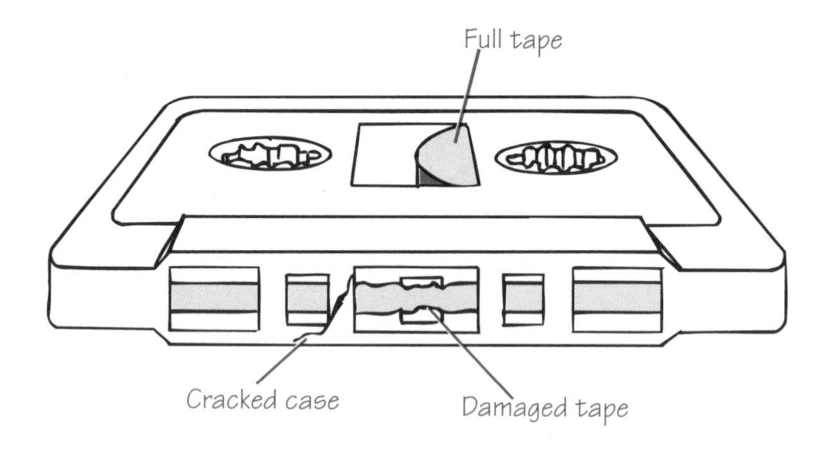

Step 9-6. Cleaning the heads.
The heads play back and record. Over time deposits can cause trouble for both functions. The heads should be cleaned on a regular basis. How frequently you clean them depends on the environment and on how often the machine is in use.

Use a swab and a cleaning solution. Technical grade isopropyl alcohol does a good cleaning job, but it should not be used on any rubber parts. For any rubber parts, use a cleaning fluid that is meant for rubber. The label will tell you.

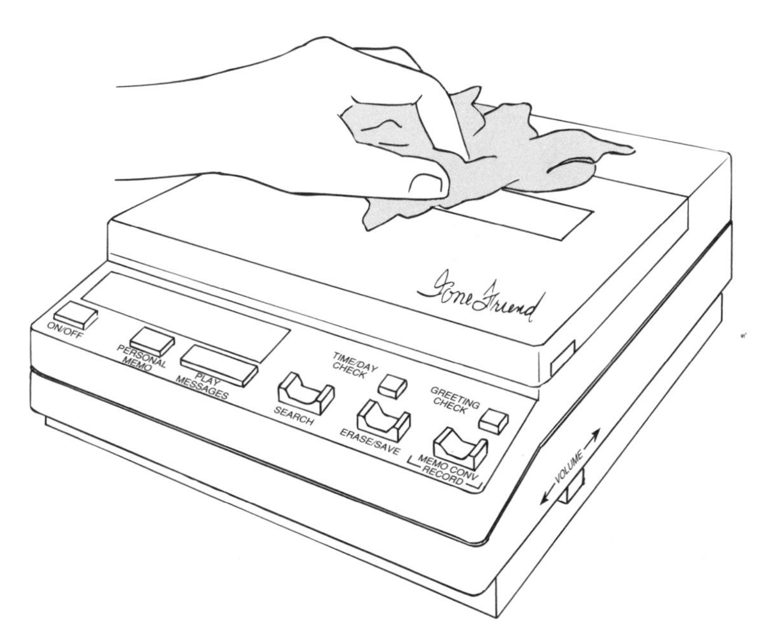

Step 9-7. Cleaning in general.
Dust, dirt and other contaminants can also build up inside the answering machine. Just as the movement of the tape over the heads will leave deposits, movement of the tape over the rollers and other parts will leave deposits.

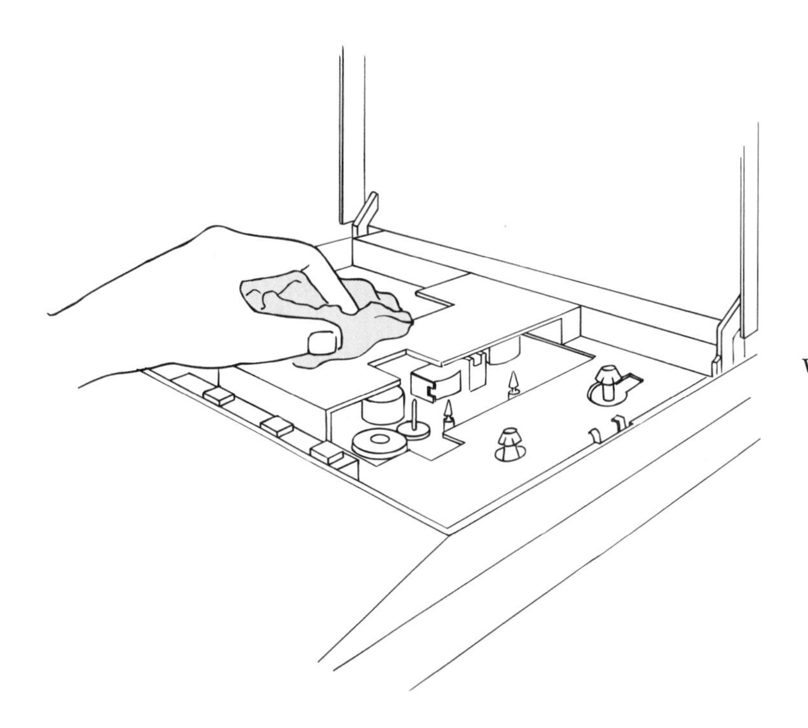

Wipe the outside with a dampened cloth.

A swab dampened slightly with technical isopropyl alcohol will work fine, but remember to keep it away from any rubber parts.

To clean the rollers, use a swab dipped in a cleaning fluid meant for cleaning rubber parts. This will be listed clearly on the fluid container's label.

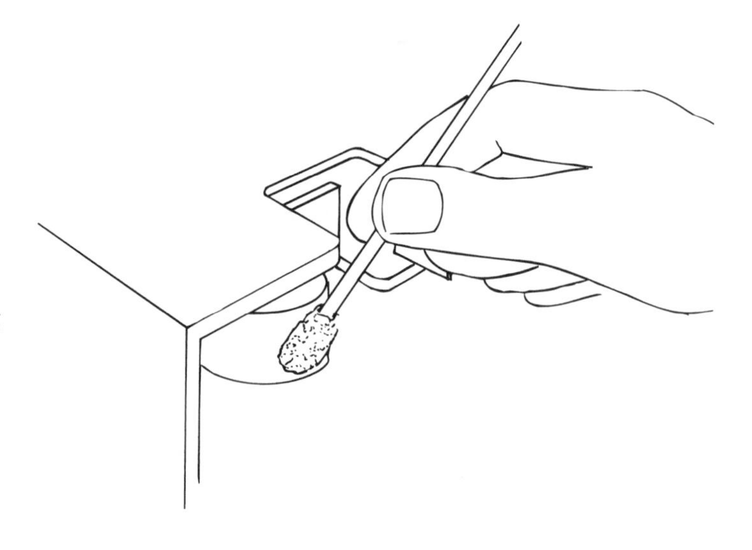

Step 9-8. Opening the case.
Most of the regular cleaning is done from the top and in the compartment where the cassettes are used. Beneath the machine are the pulleys and belts that cause the cassette tapes to move. In most cases you won't have the need to get at these very often. The answering machine should be shut off and unplugged before you attempt to open it.

The bottom of the machine is usually held in place by four Phillips-head screws, one in each corner. Remove these and set them aside in a safe place where they will not get lost. With these screws removed the bottom cover should lift off easily.

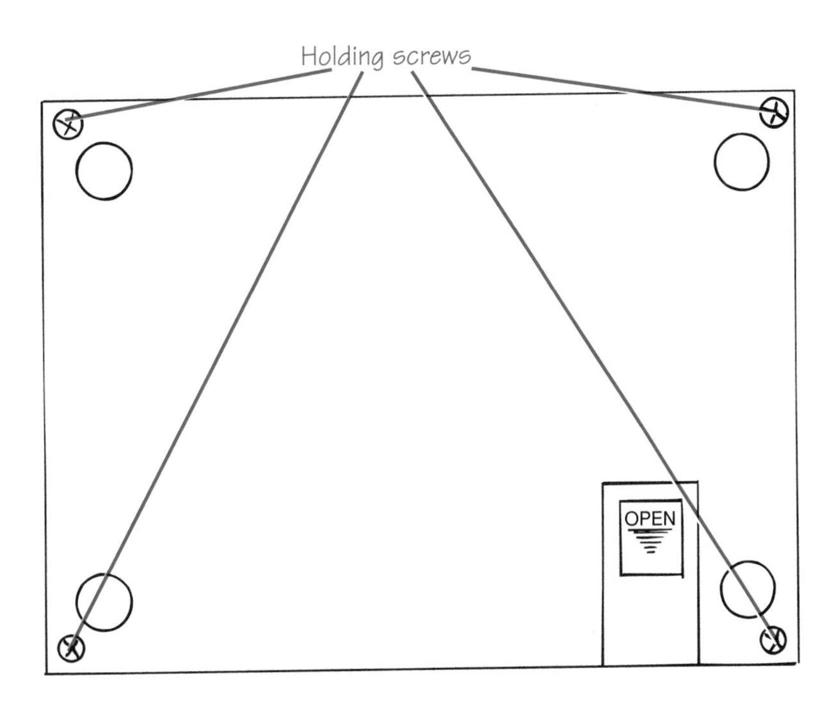

Visually examine the interior bottom. Each pulley should have a belt. If a pulley does not have a belt, chances are very good that you will also find a broken belt close by.

Check the belts for wear. One sign of this will be if the belt feels slack when you press on it (gently) with your finger.

Also examine the unit for signs of excessive dirt and dust. A little won't hurt. If there is a larger buildup, the movement might be reduced or even stopped.

In some machines, the mechanism part is attached to the top cover. In this case it will usually be connected to the main circuit board with wires and plugs. If this is how your machine is constructed, you will have to decide if it's worth going on or not. Obviously, if the machine isn't functioning, and you intend to throw it away anyway, you have nothing to lose by continuing.

If you continue, shut off the power before unplugging any connector.

Keep track of which plug goes where. Most plugs will be obvious, but others might not be. A small piece of masking tape can be used to label the wires and guide you when it comes to reinstallation.

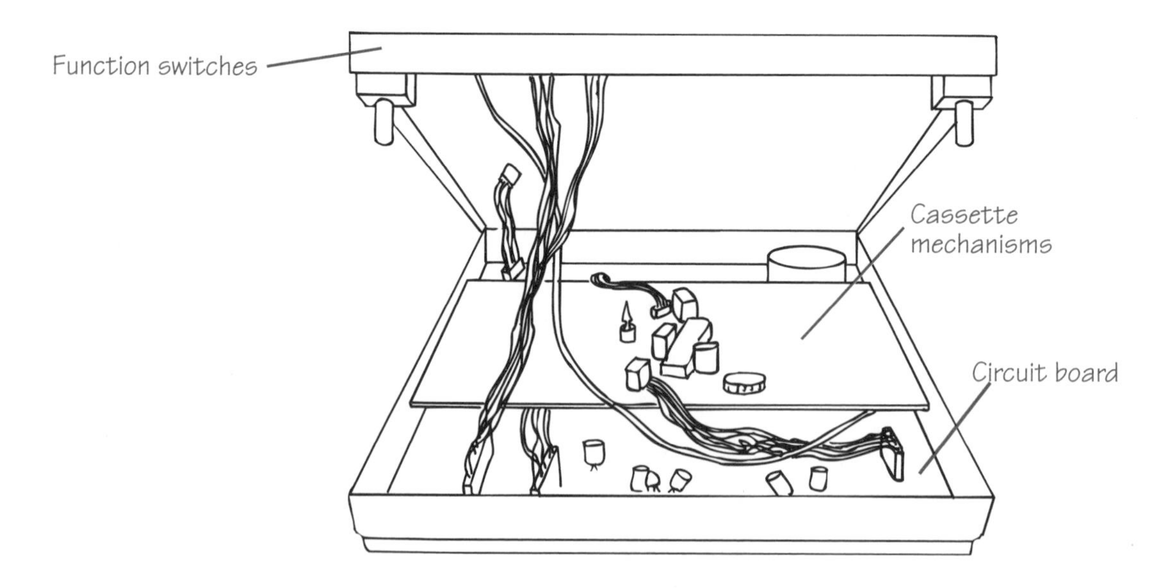

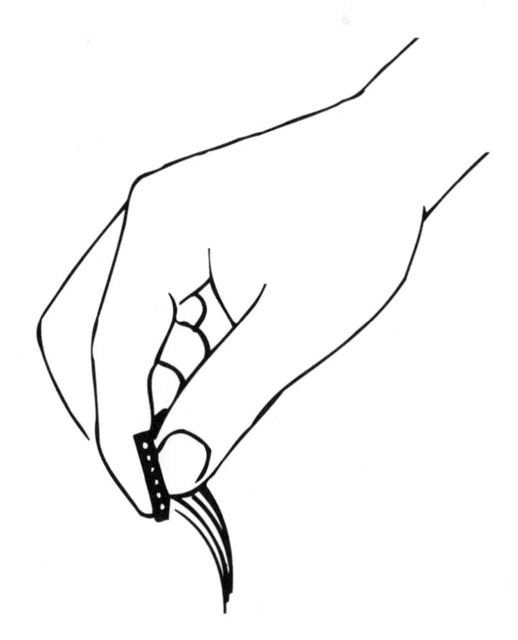

When unplugging a connector, remember that the connector and wires are delicate. Grasp the plug firmly, but not too firmly. The same is true of pulling the plug loose. Don't force it.

Once you have access to the belts, pulleys, and other mechanisms, the procedure is the same. Check for damage or wear. Clean as necessary.

APPENDIX A

How to use a VOM

Some of the tests in this book require a volt-ohmmeter (VOM, also called a multimeter). A VOM is a handy and versatile piece of testing equipment, and it's easy to use.

With a VOM, you can test any batteries you use—including the one in your car. You also can test the wall outlets to see if they have power and if they have been wired correctly (and to make sure they are safe). You can test wires, cables, and connectors to see if they need to be replaced.

The VOM you buy doesn't have to be fancy or expensive. A model costing between $10 and $20 should suffice. Extreme accuracy isn't usually needed. When checking a wall outlet for power, you rarely need accuracy of greater than about 10 percent. Even the least expensive units are more accurate than that.

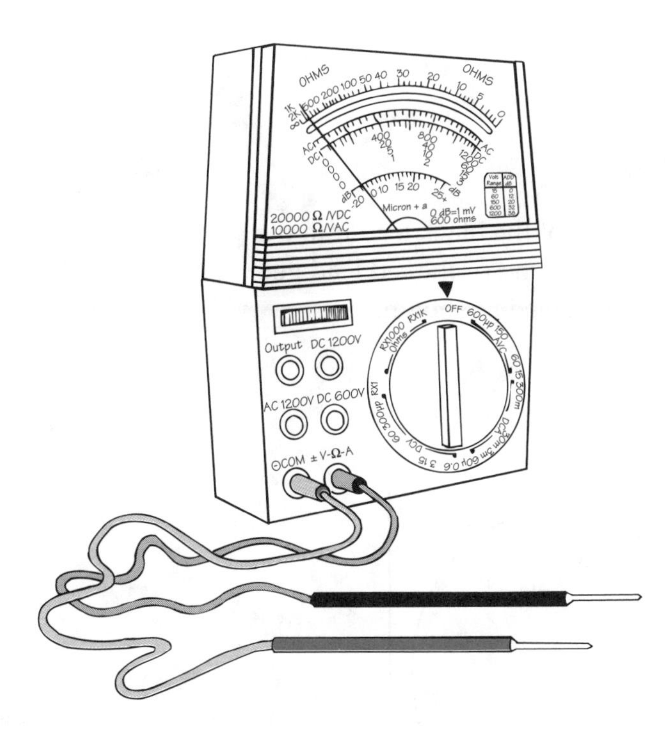

A VOM tests for ac voltage, dc voltage, and resistance. A number of ranges are available on the VOM for each test. The most common ac test checks the wall outlets in your home; your meter should have a setting for the 120-volt ac range. Common dc tests involve setting the meter in or near typical battery outputs—namely 1.5, 3, 5-6, 9, and 12. Make sure your voltage setting is higher than the voltage you are testing so you don't damage the VOM. For example, if you are expecting to read 120-volts ac, you would set the meter to the 150 setting.

If you are uncertain of the value, start at the highest setting and work downwards.

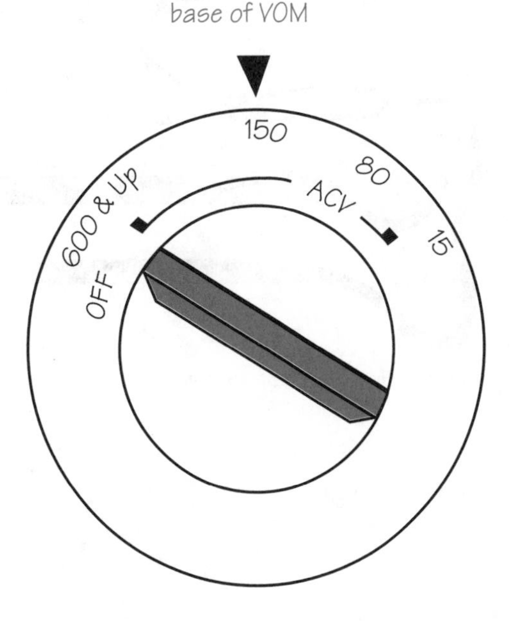

Step A-1 . Setting up the meter.
For almost every test, plug the shorter end of the black lead into the "common" (-) jack on the meter. The shorter end of the red lead gets plugged into the (+) jack, which is often labeled as shown.

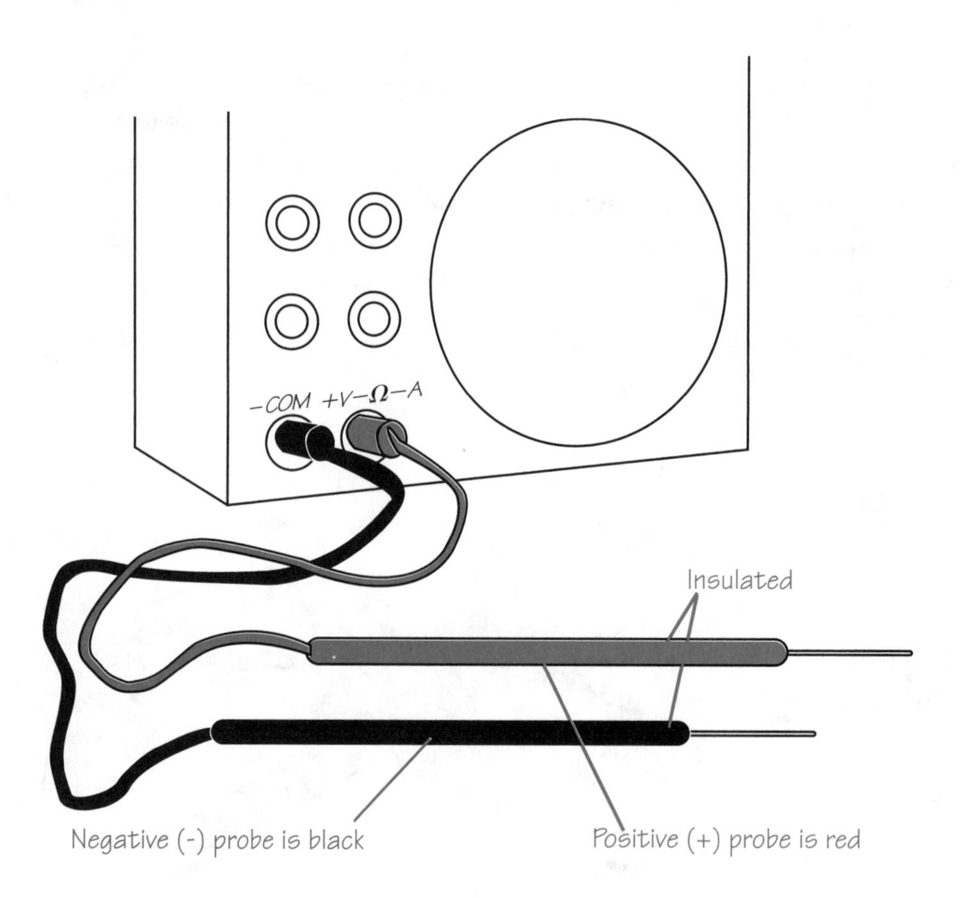

Step A-2. Setting the Ohms Adjust to zero.
To set the "Ohms Adjust," turn on the meter, and set the dial to read resistance (any range).

Touch the two probes together. Turn the adjusting wheel until the reading is exactly zero.

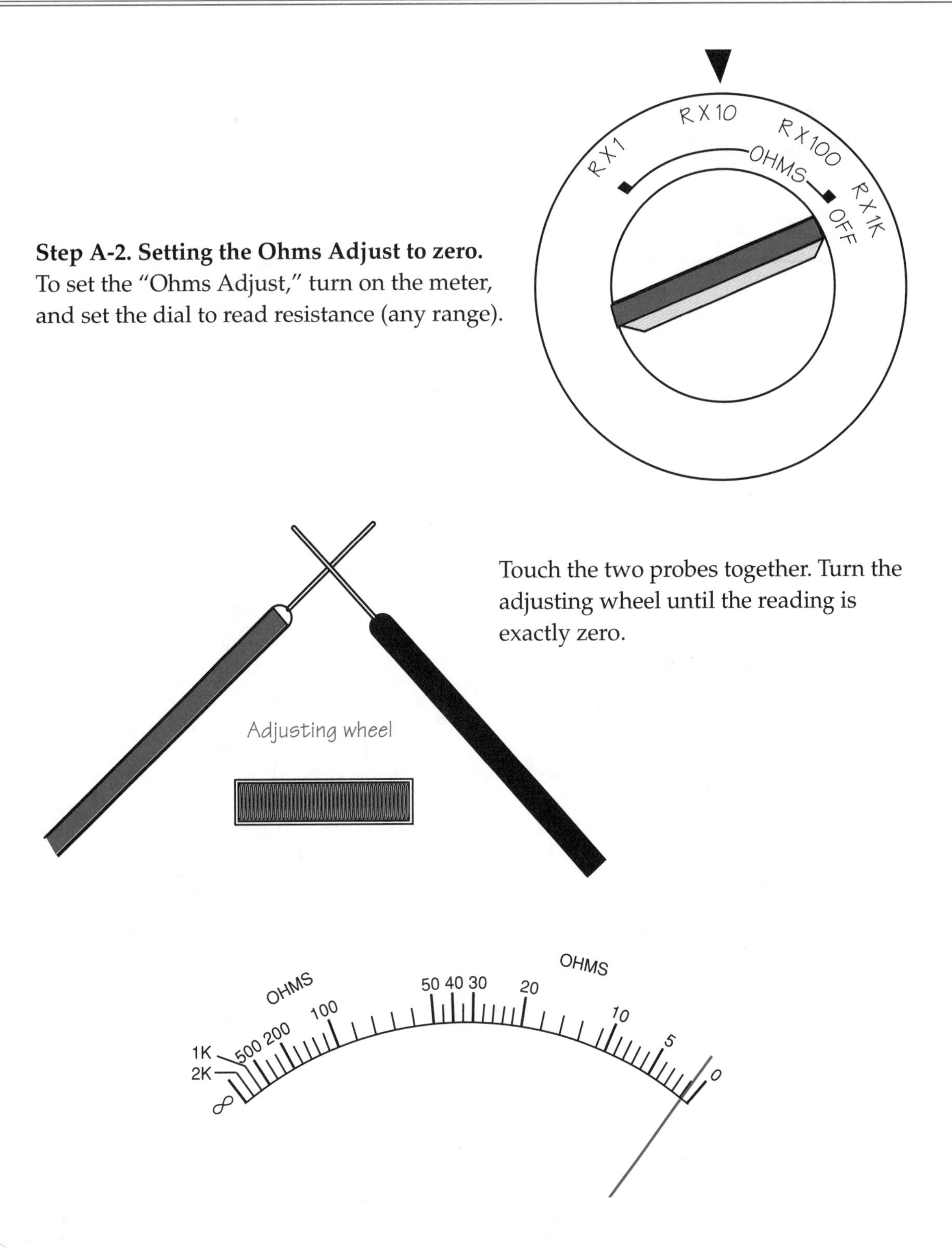

Step A-3. Setting the Zero Adjust.
If your meter has a "Zero Adjust," turn on the meter and adjust the control knob so that the meter reads exactly zero.

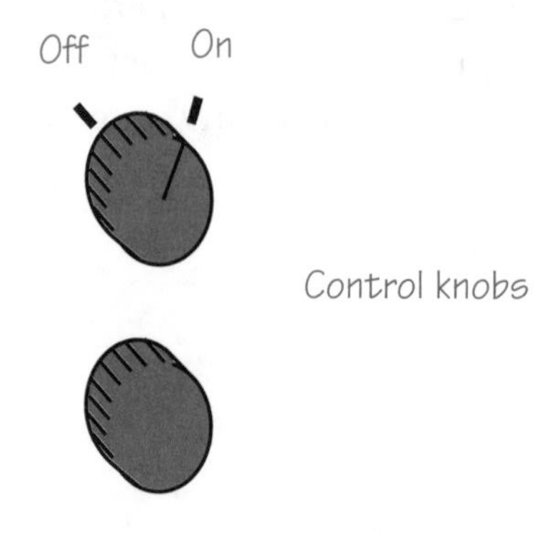

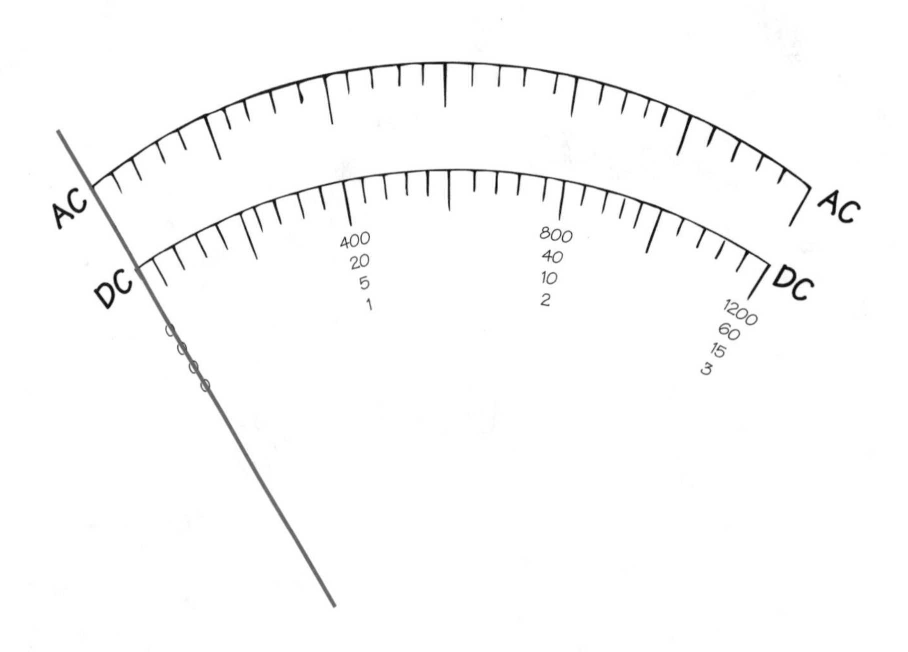

Step A-4. Reading dc volts.
To read dc volts (dc V or sometimes Vdc), set the meter to the proper range.

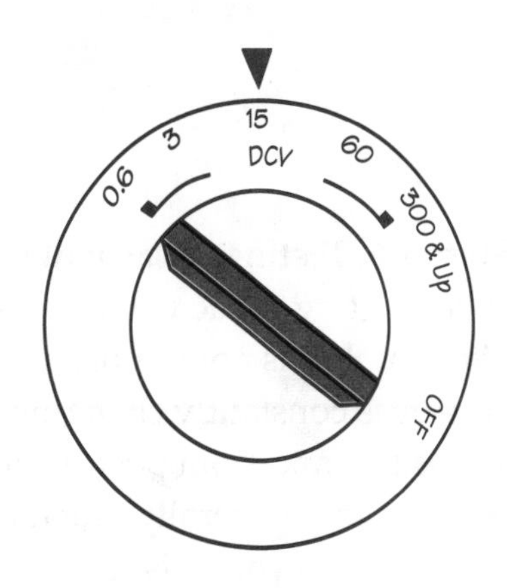

Step A-5. Testing a battery.
To test a battery, the positive terminal is usually round and is often labeled. In the case of a battery pack, the positive terminal will be labeled. Set the meter to read dc in the appropriate range. (AAA, AA, C, and D are all 1.5 volts. On most other batteries, the voltage is clearly labeled. If it's not begin with a higher setting and work downwards.)

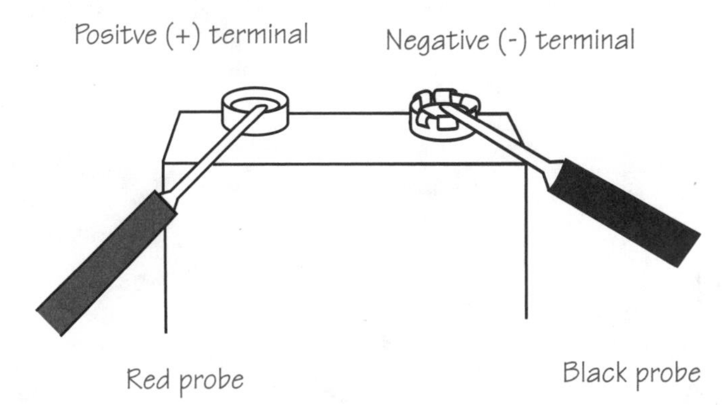

Touch the black probe to the side with the negative (– or GND) label, and touch the red probe to the side with the positive (+) label.

If the meter reads much lower than the voltage indicated on the battery, the battery is dead or is going dead.

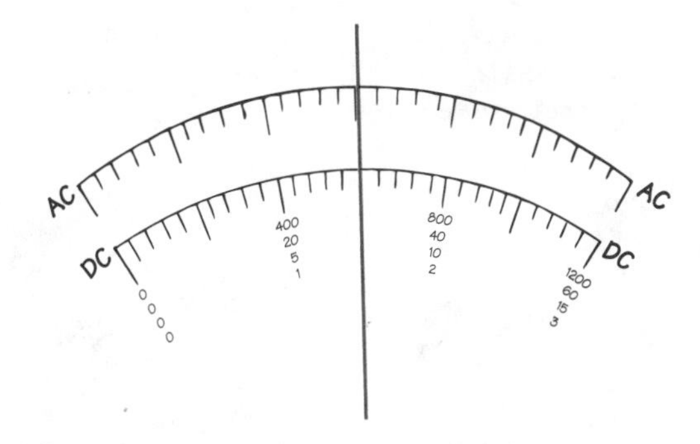

Step A-6. Testing for ac voltage.
When testing for ac voltage, the orientation of the probes is not important because the ac voltage is constantly changing from negative to positive and from positive to negative. Because ac is generally much more dangerous to you, it is essential that you hold the probes only by the insulated handles.

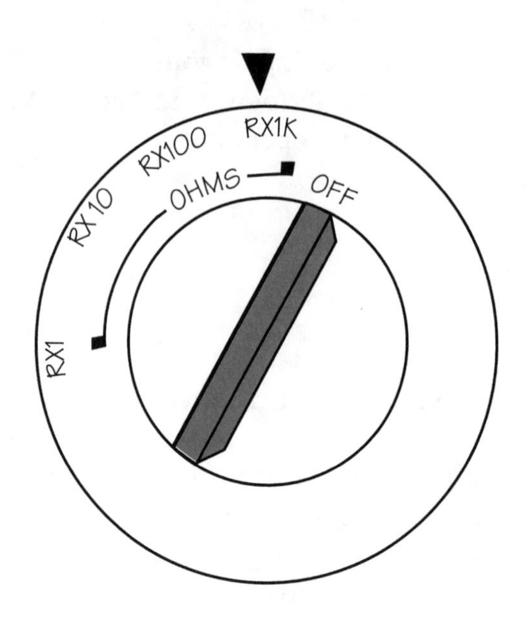

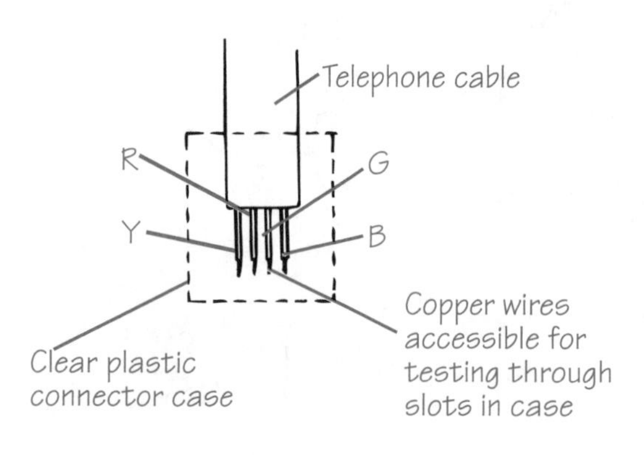

Y=Yellow
R=Red
G=Green
B=Black

Disconnect the wire or cable that is to be tested. Touch the probes to each end of the same conductor. It doesn't matter which probe touches which end. If the conductor is good, the needle should swing all the way to the right, giving a reading of zero ohms (or close to it). If there is a break in the wire, the reading will be close to infinity (full left scale and no needle movement).

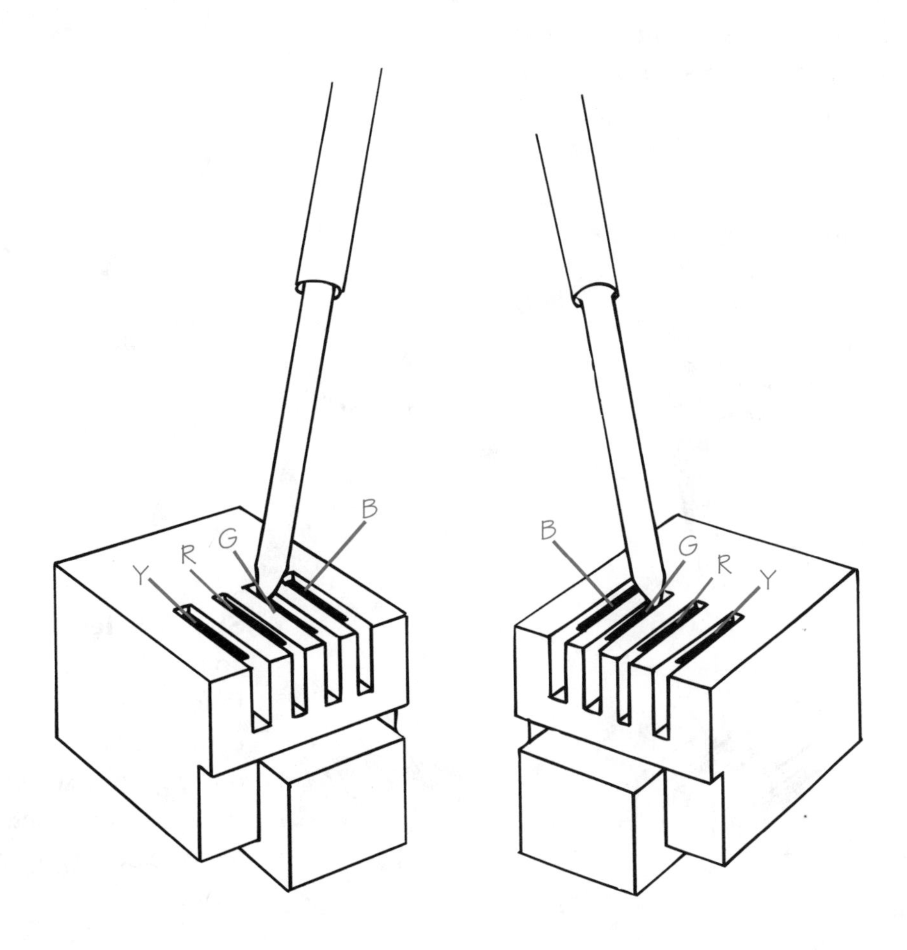

Step A-7. Testing for continuity.
In this book, the most important resistance (ohms) test is for continuity. This simply means that the conductor being tested is continuous (not broken). Although the setting used is not important, it's best to use one of the lower settings (for resistance or ohms) on the dial.

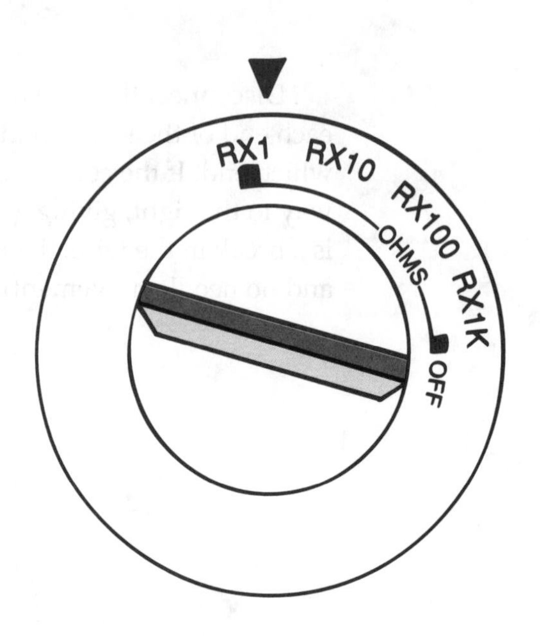

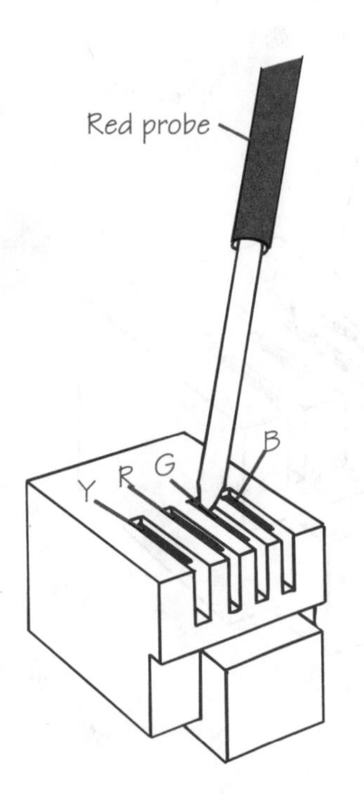

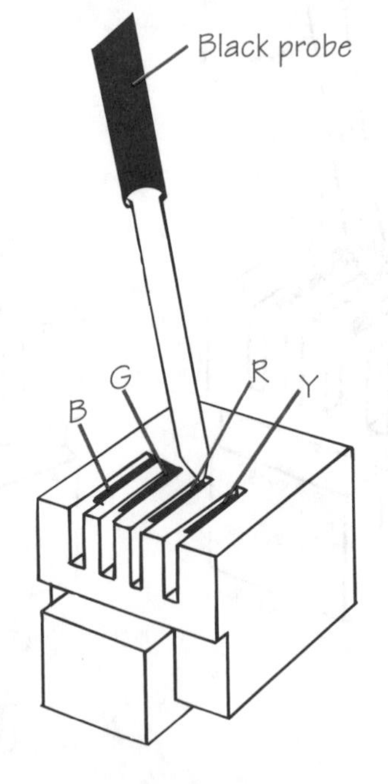

Step A-8. Testing for a short.
To test for a short in a cable, touch one probe to the center conductor and the other to the outer ground. (Repeat this using the center conductor at one end and the outer ground at the other.)

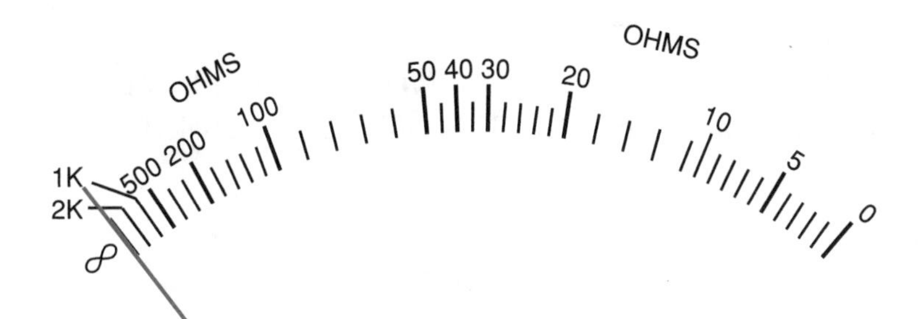

If the cable is good, the reading should be infinity (full left scale and no needle movement). If the meter gives any reading (any needle movement), the two conductors are touching.

If the meter swings all the way to zero ohms, they are in direct contact, which is a short circuit; the cable is bad.

Step A-9. Testing for dc on a circuit board.
When testing for dc on a circuit board, look for the label GND. Touch the black probe to this and the red probe to the dc point being tested. This will often be labeled something like +5 or +12.

APPENDIX B

Building a Signal Generator

Testing short cables or wires is done easily with a VOM. When the wire begins at one location and ends at some distant place, using the VOM by itself is impractical. An example of this is when you need to test the wiring that goes through the walls of the house between the access box and each of the various telephone outlets throughout the house.

This is done by generating a signal and sending it through the wires from the access box. If the signal appears at the same wires of the wall jack, those wires are intact; if the signal does not appear at the jack, the wires between are broken or are otherwise damaged.

Professional telephone repair technicians have some special equipment for this kind of testing. Although you can find and purchase your own equipment, the cost is usually prohibitive. However, you can very easily and simply make your own.

It should be noted that the telephone is using only one pair of wires, usually colored red and green. See chapters 5 and 6 for more information on this. The goal of this appendix is merely to show you how to build a simple signal generator and basically how to use it. If the wires in the four-wire cable are found to be broken, you can simply use the other color pair. Before you can do this, however, you have to find out which wires are good and which are bad.

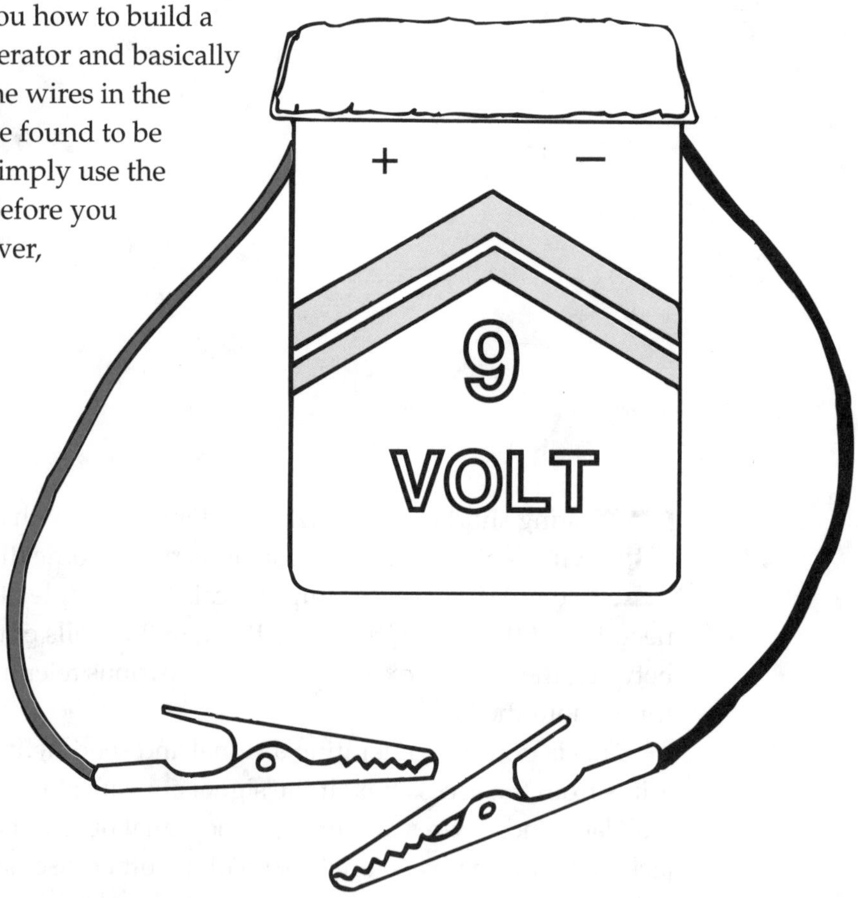

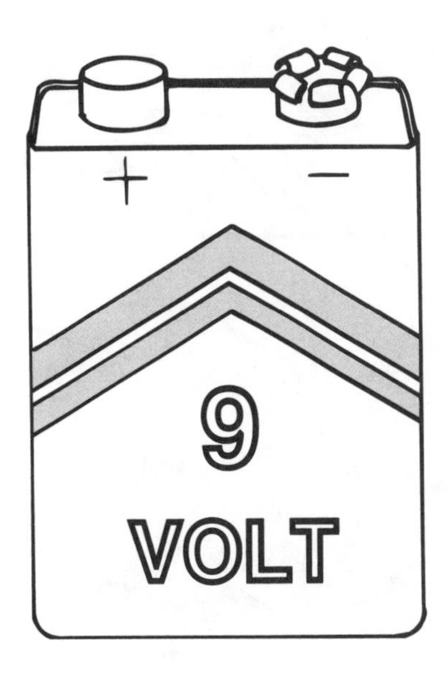

The easiest way to find out is to use a battery to send voltage through the wires. Although just about any battery can be used, a 9-volt battery and connector makes things very easy.

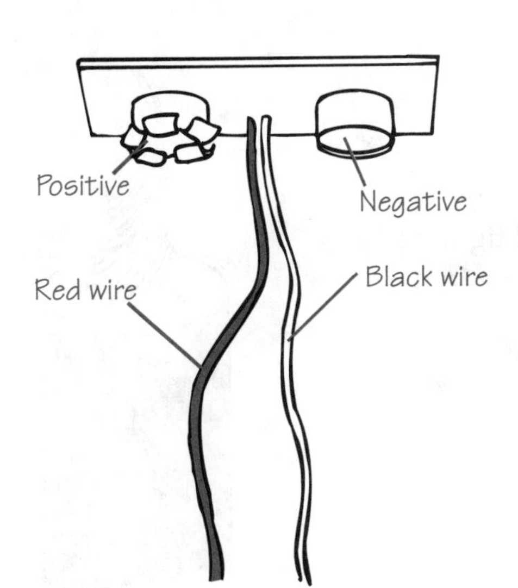

The connector for a 9-volt battery simply snaps onto the top of the battery. It comes complete with two wires. Most often these are already bared and may also be "tinned" (coated with solder).

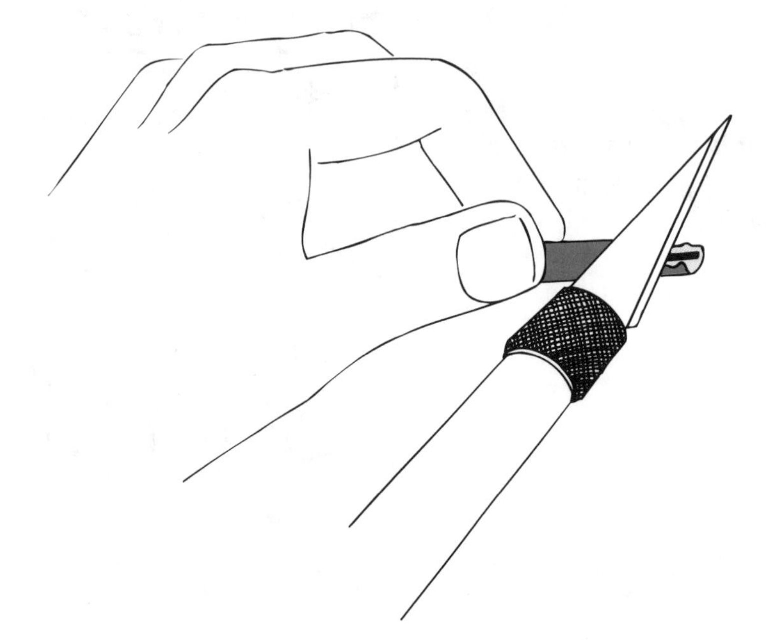

Stripping the wires

If the wires are not bare, use a sharp knife or wire strippers to remove about ¼ inch of the insulation. Be very careful. Not only might you cut yourself, the stranded wires inside are small and can be easily damaged.

Twisting the wires together

With the insulation removed, use your fingers and tightly twist the wires together. (If you know how, use a soldering iron and give the wire ends a light coating of solder.)

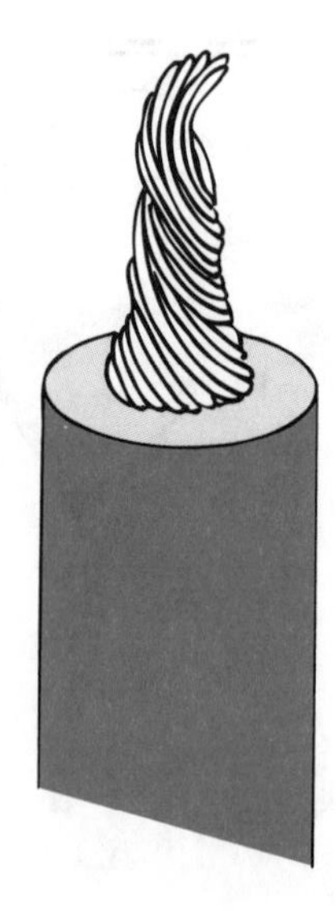

Top view of clip

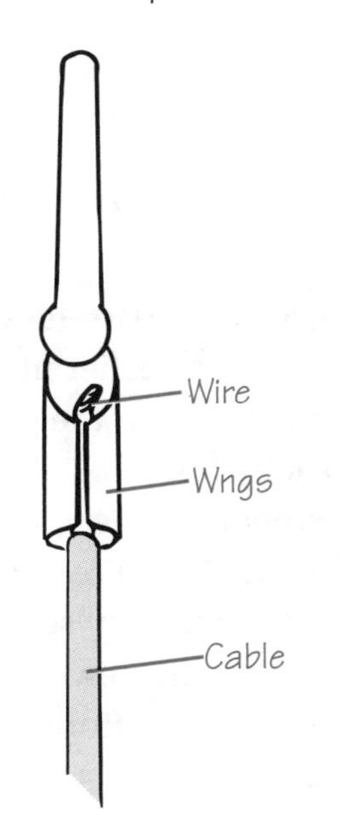

Inserting the wire into the clip.

Insert the wire into the clamp-end of an alligator clip. This can be a round hole, or the clip might have "wings" on each side. Whichever is used, the clamping part should be suitable for the wire size. (If you don't intend to solder, be sure to get solderless-type clips.)

Side view of clip

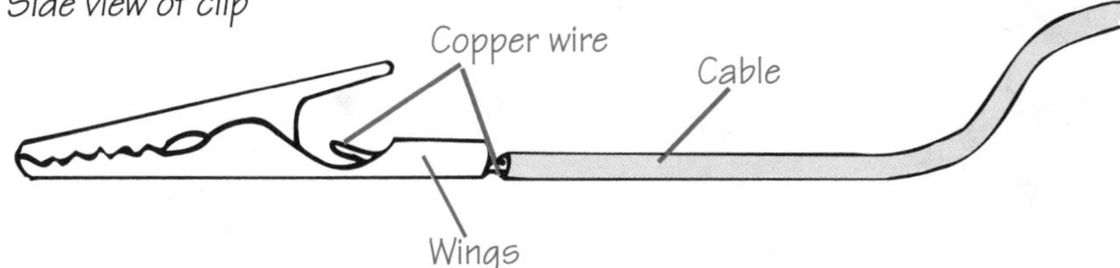

Crimping the clip

Using a pair of pliers or a crimping tool, crimp the connector end onto the wire. Especially if you are not going to solder, it is important that this connection is tight and secure.

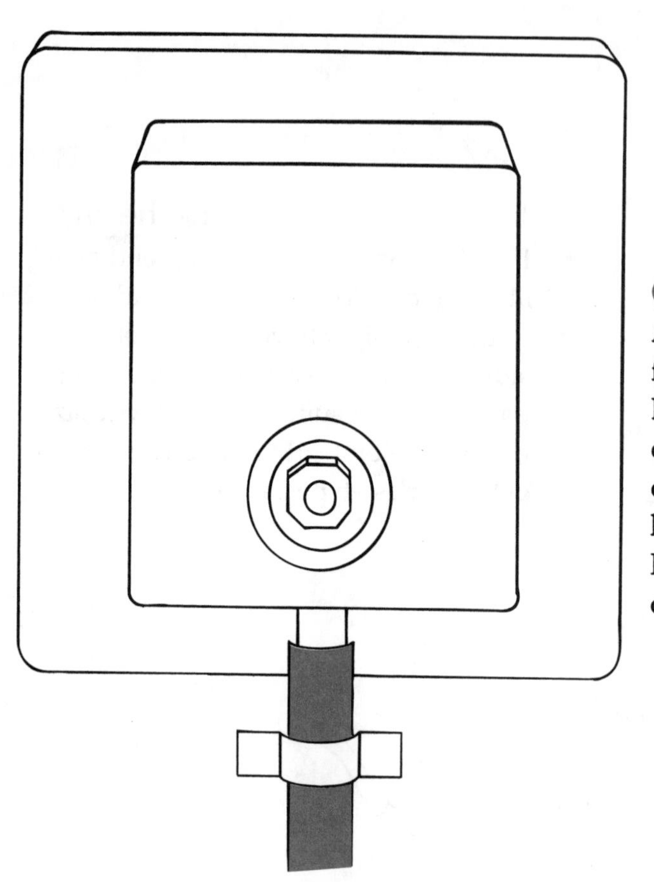

Opening the access box

As shown earlier in this book, find and open the access box. Exactly how this is done depends on the way your box is designed. The cover might be held in place by screws or bolts. Many of the newer ones have a clip on the bottom.

Side view of access terminal

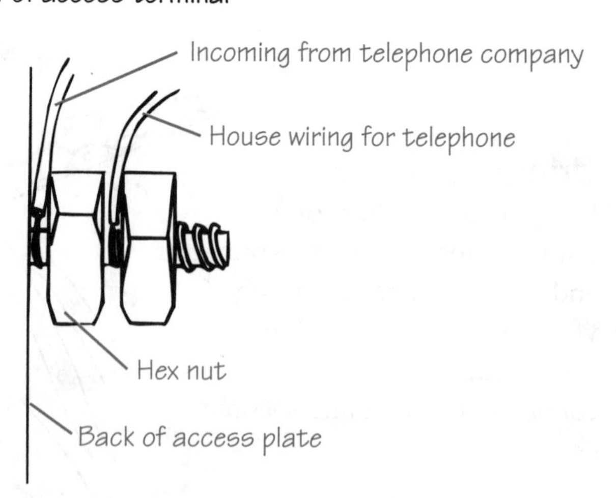

Disconnecting the house wires

Most often the telephone company's cable are larger, with the wires connected to the bottom parts of the lugs, and the household wires are smaller and connected to the top. Find the wires that go into the house and carefully disconnect them.

There are two important precautions. One is to be sure that the color coding used is visible. (You should be able to see at least some of the colored insulation.) If it's not visible, label the lugs so that you know how to reconnect the wires later. The second precaution is to be sure that you are wearing rubber soled shoes. It's unlikely, *but possible*, to get a shock from the wires.

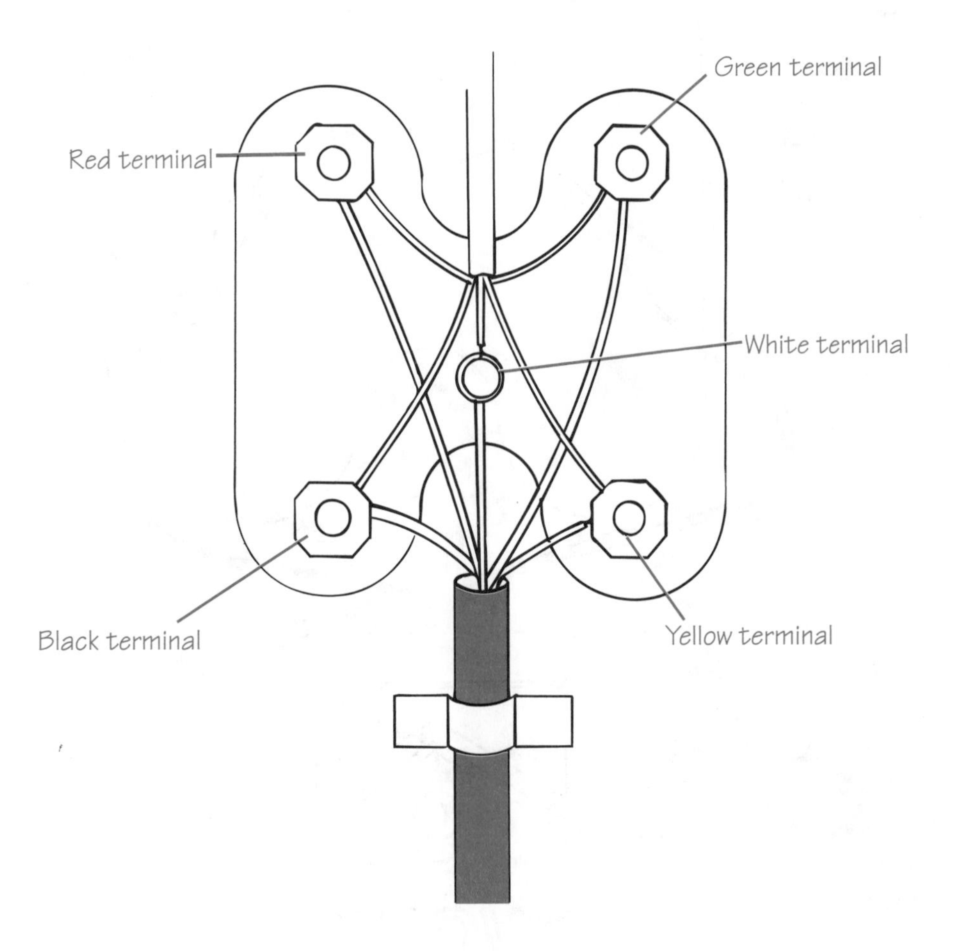

Attaching the alligator clips to the wires.

Attach the alligator clips from the generator to two of the wires that go into the home. You will probably begin with the red and green wires. Note which wire is connected to the positive side of the battery and which is connected to the negative side.

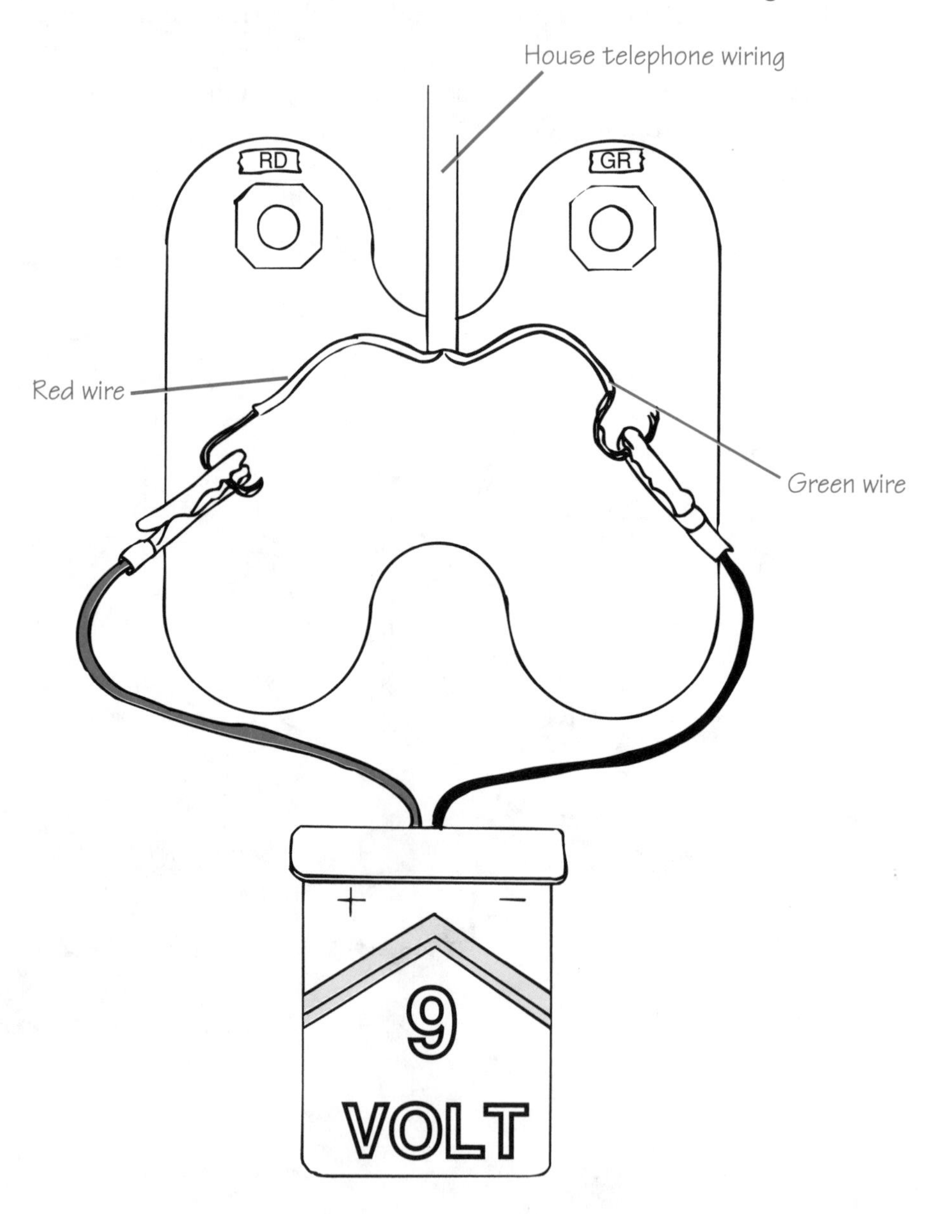

Removing jack plate

If need be, remove the jack plate and turn it over.

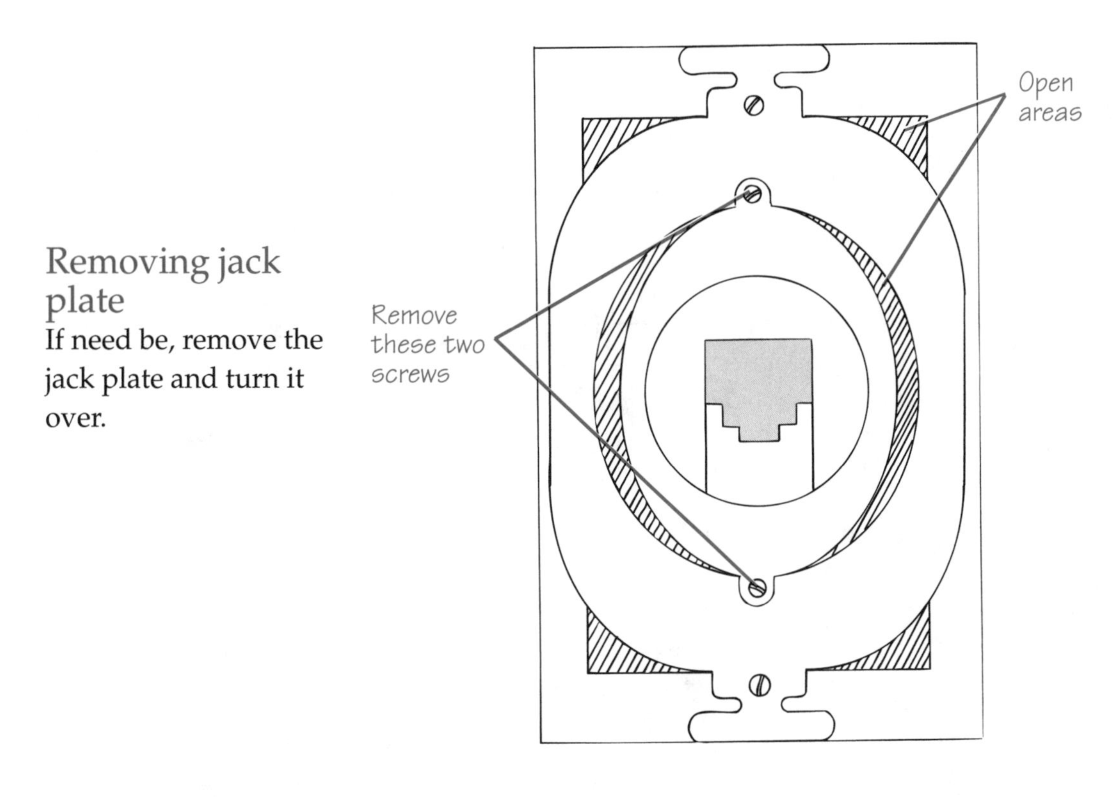

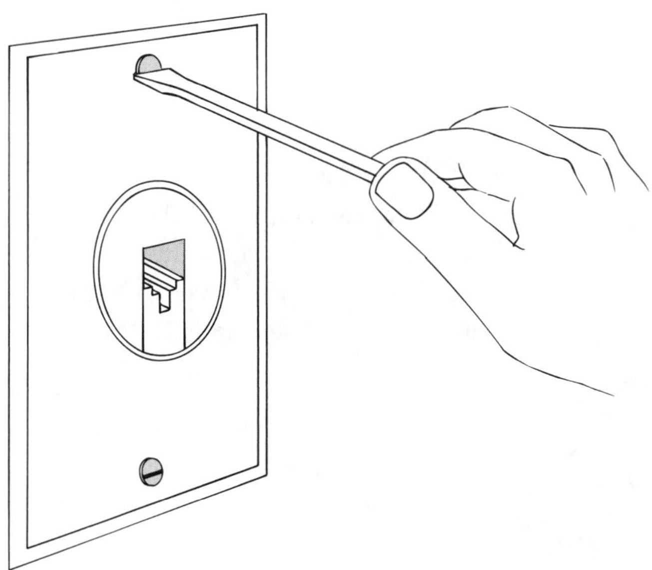

Removing the wall plate inside

Inside your home, remove the wall plate of the suspected jack.

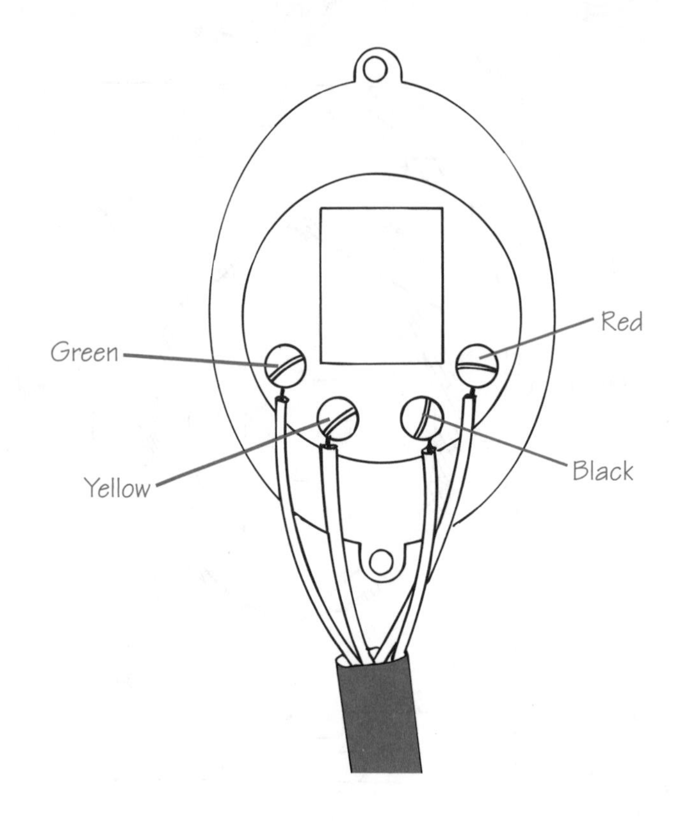

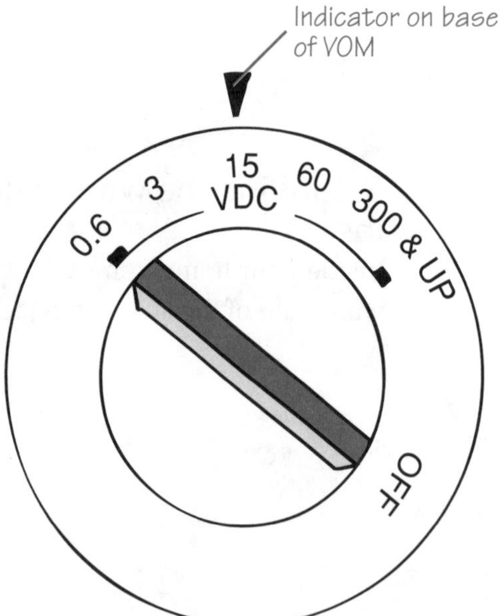

Testing with the VOM

Set the VOM to read dc volts. The setting should be the nearest to the battery voltage (9 volts in this case) and above it. On the VOM used here, that setting is 15 Vdc.

Pay attention to the polarity (+ and –), probe the appropriate terminals of the wall jack, and read the meter.

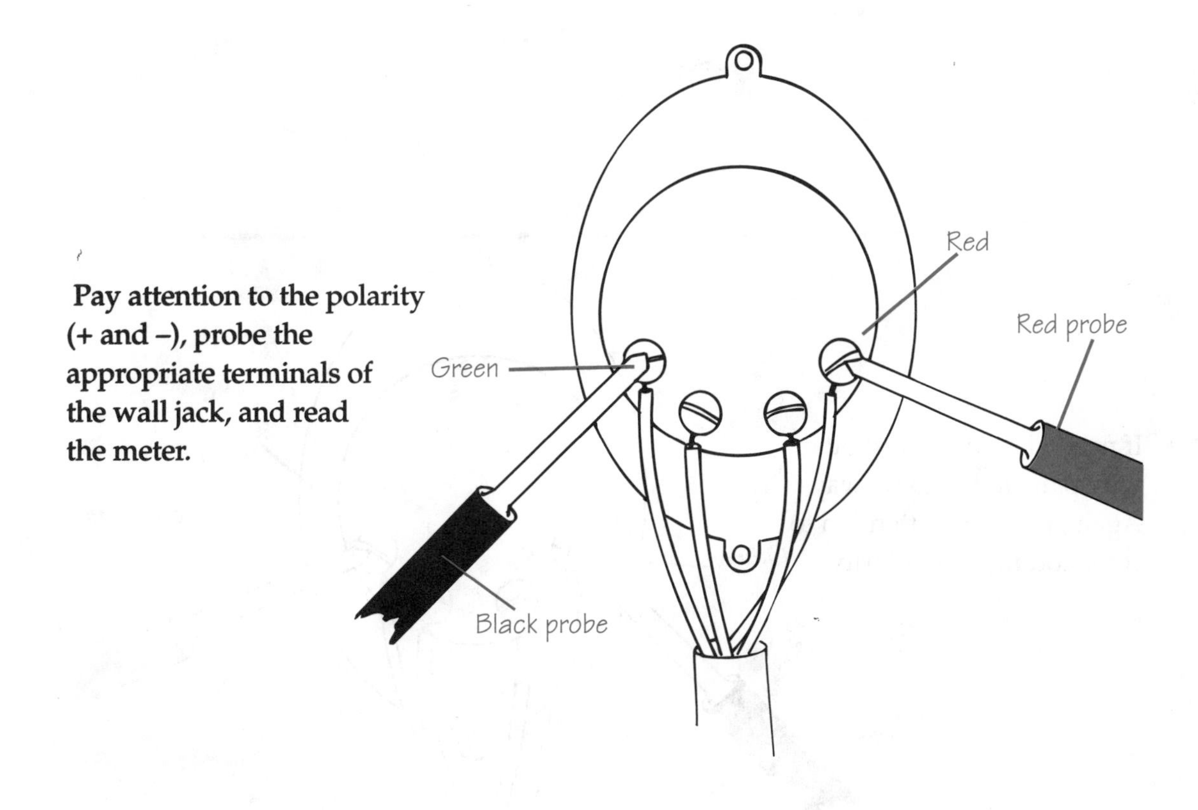

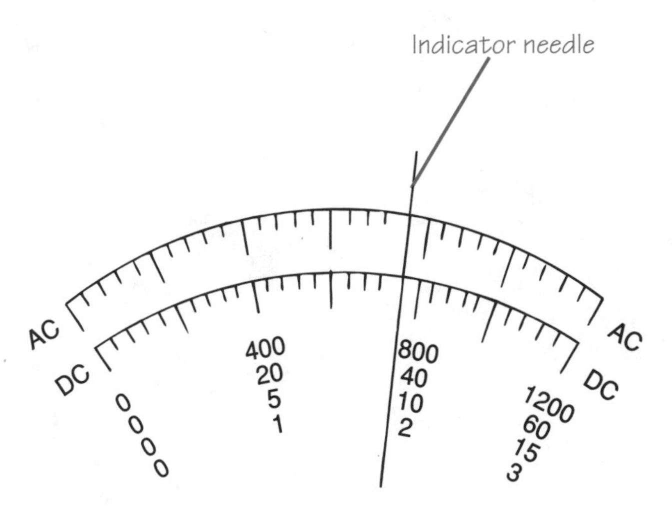

Assuming that the battery is good, the reading you get should be very near that voltage. If you get no reading, either the battery is dead, you are touching the wrong terminals, or at least one of those wires is broken.

If needed, you can test other wire pairs in the same way. Again, pay close attention to color coding and polarity.

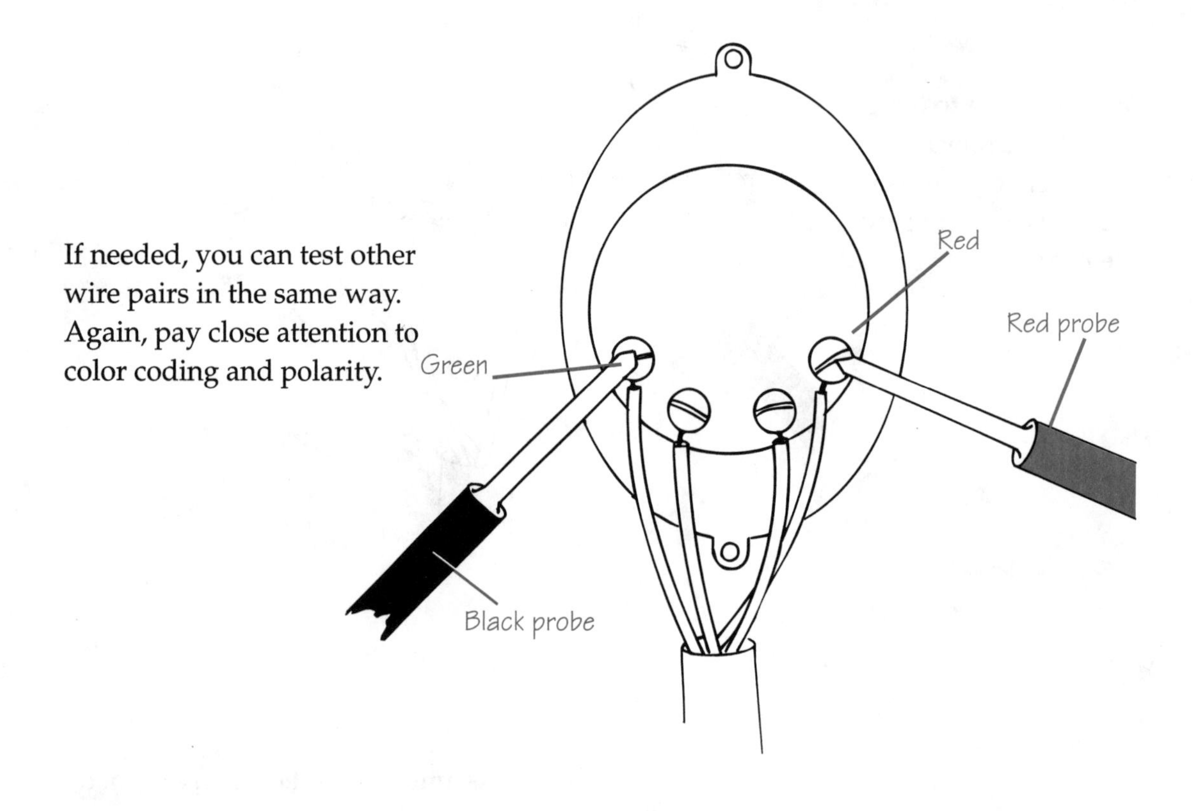

Glossary

ac
Alternating current, such as the power from a household electrical outlet. The value is constantly changing in a sine wave.

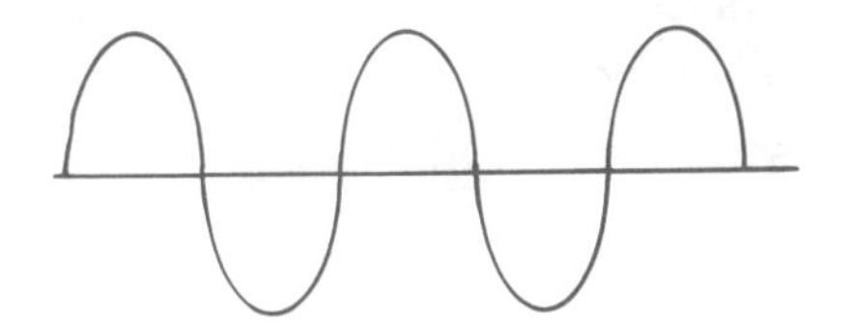

access box Also called the "service box" and sometimes "house protector." This is where the phone company's lines enter the home.

adapter
A device to change the incoming 120 Vac into a needed value of dc.

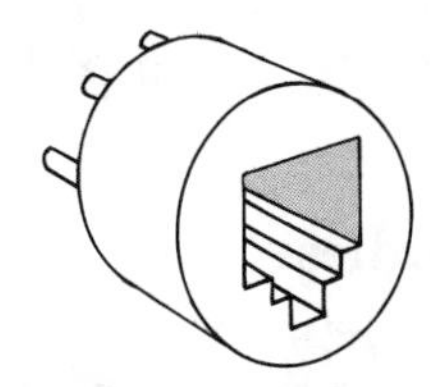

base The part of the telephone that is more-or-less stationary and connected to the main telephone lines. (See also *handset*.)

battery
An electrochemical device that generates electricity through a chemical reaction. For telephones, batteries are used to back up internal memory if the main power is interrupted and to power cordless phones.

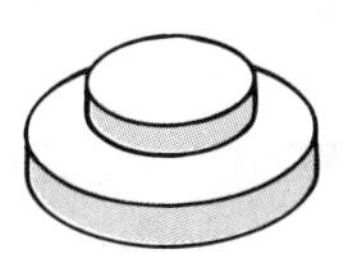

battery pack
More than one battery cell hooked together inside a single package.

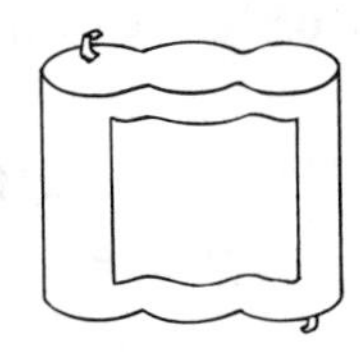

cable
A combination of two or more insulated conductors in a single casing.

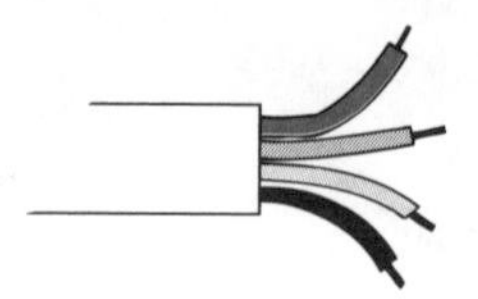

cassette
The package holding recording tape for an answering machine. Two sizes are available. The "standard size" is the same as a regular audio cassette. The microcassette is much smaller.

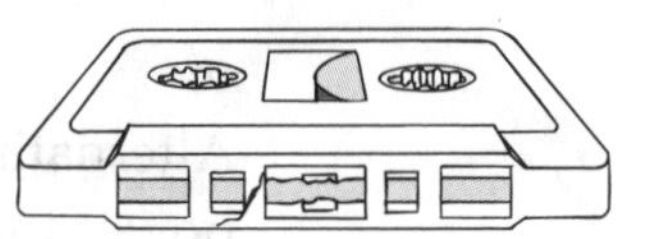

cellular A communications system which uses interlinked receivers/transmitters, each of which covers a particular area, or cell. Used for mobile and other portable telephone communications.

circuit A complete electrical path.

cordless
A telephone that uses radio waves to send signals back and forth between the handset and the base. The base is connected to the phone lines.

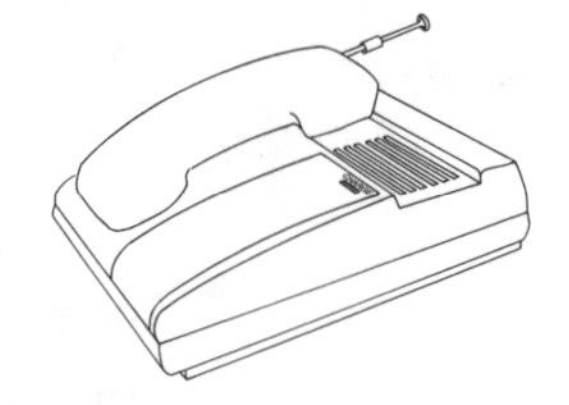

dc Direct current, such as that in a battery, or provided to the computer circuits by the power supply.

DTMF Dual tone modulating frequencies. The method by which a touchpad dials.

duplex Communications that go in both directions at the same time.

FCC Federal Communications Commission. The governmental group that oversees communications in America, particularly when it concerns radio waves.

ground The negative or return path. Also used to describe the ground itself, where an excess voltage can be safely drained in the event of a short circuit.

handset The part of the telephone you hold in your hand.

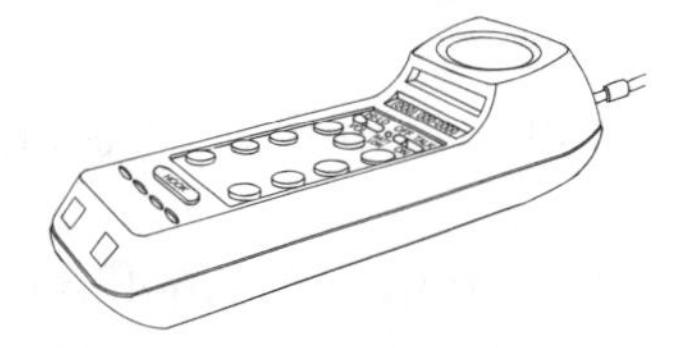

hook switch The switch which is automatically activated when the handset is placed into the cradle, breaking the connection.

modular The smaller connector or jack now used for most telephone installations.

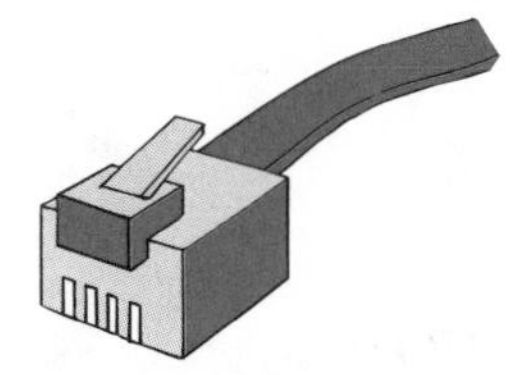

nicad Nickel-cadmium. A type of rechargeable battery.

pedestal The box often used by the phone company as the junction box for a number of local telephones.

polarity The + and – of electrical current flow. When testing, the red probe of the VOM goes to the + and the black goes to the – or ground.

pulse dialing The method of dialing by which electrical pulses (ons and offs) are "counted" by the phone company's switching computers. See also *rotary dialing*.

rotary dialing The pulse method of dialing in which a rotary dial causes a series of ons and offs to be sent. These are "counted" and the connection is made.

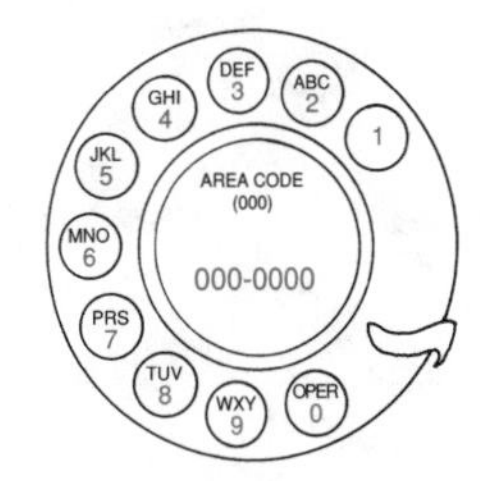

simplex Communication in one direction at a time only.

switching station The building and computerized equipment that routes calls.

ohm The unit of resistance to the flow of electrical current.

touch tone The method of dialing which uses generated tones instead of pulses (see *DTMF*). As a key is pressed, two tones are generated simultaneously—one from the horizontal line and one from the vertical. The combination makes the specific tone that represents that particular key.

VOM
Volt-ohmmeter, also called a multimeter because it can measure voltage, resistance (in ohms) and often current (in milliamps).

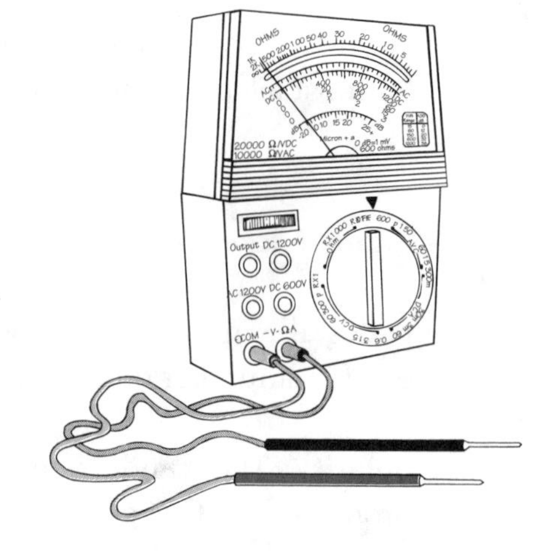

Index